THE
STAR GUIDE

THE
STAR GUIDE

**A Unique System
for Identifying
the Brightest Stars
in the Night Sky**

BY
Steven L. Beyer

**With Maps And Illustrations
By The Author**

Little, Brown and Company
Boston   Toronto   London

Library of Congress Cataloging-in-Publication Data

Beyer, Steven L. (Steven Larsen)
   The star guide.

   Bibliography: p.
   Includes index.
   1. Astronomy — Amateurs' manuals.   2. Stars —
Amateurs' manuals.   I. Title.
QB63.B49   1985   523.8   85-13039
ISBN 0-316-09267-3
ISBN 0-316-09268-1 (pbk.)
HC: 10   9   8   7   6   5   4   3   2   1
PB: 10   9   8   7   6   5   4   3

BP

DESIGNED BY DEDE CUMMINGS

*Published simultaneously in Canada
by Little, Brown & Company (Canada) Limited*

PRINTED IN THE UNITED STATES OF AMERICA

The heavens declare the glory of God,
and the expansion shows his handiwork.
Daily they speak, and
nightly they display knowledge.

Psalm 19:1 – 2

He who has not contemplated the mind
of nature which is said to exist in
the stars . . . is not able to give a
reason of such things as have a reason.

Plato, Laws XII

Contents

Acknowledgments

Many people have directly or indirectly helped me with the preparation of this book. I especially want to thank Andrea Elgin Beyer for her inspiration and many creative contributions to the development of The Star Guide.

I would like to express my appreciation to Professor Norman H. Baker of Columbia University, for many insights learned in his graduate course, Basic Astronomical Data. Kenneth L. Franklin, Astronomer at the American Museum–Hayden Planetarium, provided advice that was of great help. Allen Seltzer also helped in many ways. George Lovi gave expert suggestions about the presentation of data and the construction of seasonal star maps. Professor Frederick H. Willecke of Wagner College provided translation and interpretation of material related to the origins of star names. Joseph Patterson of the Department of Astronomy, Columbia University, advised on a number of subjects. Sandra Kitt arranged for the use of the Perkins Memorial Library of the American Museum–Hayden Planetarium.

I also thank Professors Willard Jacobson, Warren Yasso, O. Roger Anderson, and Paul Rosenbloom of the Department of Mathematics and Science Education of Teachers College, Columbia University; Professor Rodney Doran of the State University of New York at Buffalo; and Professor Samuel Devons and George Tremberger, Jr., of Columbia University, for their insights and information.

The American Museum of Natural History and the American Museum–Hayden Planetarium have long provided inspiration for me, and I thank Museum Director Thomas Nichol-

son and Planetarium Chairman William Gutsch for the privilege of teaching at the Planetarium. Sky Show producer Clarence Brown has been most helpful in a number of areas, as was Thomas Lesser. Planetarium technicians Joseph Maddi, Joseph Doti, Kai Eng, and Julio Marrero have enabled me to efficiently use the Zeiss projector and other equipment. Lynne Azarchi, Francine Oliver, David Roth, Violet Pena, and David Ross have given assistance. Fred Hess, Henry Bartol, and Samuel Storch have also been helpful with information and advice.

Some of those who aided in the assembly of materials were Agnes Paulsen of Kitt Peak National Observatory; Robert Kraft, Nancy Hanson, and Pat Shand of Lick Observatory; Margaret Weems of the National Radio Astronomy Observatory; William Morgan and Richard Dreiser of Yerkes Observatory; Kathy Ronan of the Perkin-Elmer Corporation; and Dorotha Peltier.

Tom Shidemantle provided his skills in photographic reproduction. Michael Conroy and Vincent Bellafiore were helpful with advice and suggestions.

My thanks also to author Robert Burnham, Jr., and Mary Duffy of Dover Publications for the quotation from Burnham's Celestial Handbook. Thanks also to author Berton Willard and Jeanne Brown of the Cumberland Press for the passage from Russell W. Porter: Arctic Explorer, Artist, Telescope Maker.

It is my hope that The Star Guide will help you share the joy that stargazing has given me, and I want to thank some of those who took time to help me learn about the heavens.

My parents, Sophie Larsen and Svend Aage Beyer, actively encouraged me to appreciate and study the wonders of our world and sky. Mary Larsen, Laurentzius Beyer, Astrid Rock, Barbara Nielsen, Louis Beyer, Asta Jackson, Kenneth Jackson, Roy Downing, Vincent Gattullo, and many others also helped me acquire a love of learning. Robert Engler was among the first to point out constellations to me. Richard Luce, a neighbor and Director of the Optical Division of the Amateur Astronomers Association in New York, showed me stars and planets through a telescope that he had built. From his son, Paul Luce, I learned about making accurate sky observations. Patrick Rizzo enthusiastically encouraged my astronomical interest, as did Hans Behm.

Franklin Branley, Mark Chartrand, and Tom Carey invited me to teach at the Hayden Planetarium. Marcy and Phil Sigler gave their interest and enthusiasm. George Chaplenko, Frank Biribauer, Alan Witzgall, and other members of Amateur Astronomers, Inc., of Cranford, New Jersey, graciously introduced me to facilities of the William Miller Sperry Observatory at Union County College.

I also give thanks to my students, whose enthusiasm, questions, and suggestions have helped me to prepare The Star Guide.

In addition, I wish to express my admiration for astronomers who have contributed to our knowledge of the universe. The individuals mentioned in this book are representative of many others who have worked diligently and creatively in the study of the heavens.

Appreciation and thanks are given to Little, Brown and Company and to my editor, Christina H. Coffin, for her advice, encouragement, professional skills, and patience. I am also extremely grateful to Dede Cummings for her work in developing The Star Guide's design and to copyeditors Peggy Freudenthal and Doris Heitmann for their diligence.

Although I have had help in the preparation of this book, I take full responsibility for its content and would appreciate hearing from you about mistakes that may have escaped my detection.

Introduction

STARS appear as beautiful sparkling points of light, and their lofty presence inspires us to learn more about them. The night sky is one of nature's visual treasures as well as a manifestation of our ultimate origin and destiny. My hope is that using *The Star Guide* will increase your appreciation of the sky's magnificence.

We are the progeny of stars in the sense that most chemical elements in our bodies were forged during the evolution of ancient stars. Matter from those stars later coalesced to form the earth, and for approximately 4.5 billion years the sun, our nearest star, has provided the energy that animates life on this planet. During this time it is likely that various forms of celestial energy, including ultraviolet light and cosmic rays, helped alter the course of biological evolution by affecting genetic material in plants and animals.

During the Devonian Period of earth's natural history, about 400 million years ago, some life forms began a migration from the seas to the land. Today, we find ourselves on the threshold of a far wider sea, whose ultimate challenges and rewards cannot yet be imagined. As we gaze at the stars from a sidewalk or backyard, we look towards a realm of space and time that beckons us to life's next arena in a majestic passage through the ages.

Man now dominates other forms of life on earth. We challenge each other and the deepest mysteries of nature. At this stage, we may indulge in the luxury of savoring our natural world, looking at the stars, and dreaming of our origins. As you watch the stars or listen to the thunder of ocean waves, you may sense our distant heritage.

It is said that Beethoven was inspired to compose the second movement of his eighth quartet, Opus 59 No. 2 in E Minor, while stargazing on a summer night from the town of Baden near Vienna. Beethoven noted that "this piece is to be played with much feeling," and the music has been described as being evocative of emotions we may feel when inspired by the glory of the night sky. Vincent van Gogh was another who was inspired by the stars, as may be seen in his paintings *The Starry Night* and *Starry Night on the Rhône.*

In addition to the sublime visual aspect of the stars, astronomers have learned much about their physical characteristics. *The Star Guide* will introduce you to many of the brightest stars so that they will become familiar and can evoke fond memories of the times when you learned their names and how to identify them. The ability to recognize stars is a skill that is a pleasure to enjoy alone or to share with friends.

The Scottish essayist and historian Thomas Carlyle once wrote: "Why did no one teach me the constellations when I was a child?" This lament need not apply to you. Use *The Star Guide* to make friends with the stars and in this process enhance your sense of place in the universe.

Part I
Before You Begin

The Universe

LET us briefly consider what has been learned about the origins of the universe and the stars themselves.

The universe is believed to have begun between 10 and 20 billion years ago in an explosive event known as the Big Bang. Within the primordial fireball of this Big Bang, basic components of our universe were forged under conditions of enormous pressure and temperature. As the fireball expanded it cooled, and after about 700,000 years its temperature is believed to have been low enough for electrons to join with protons and neutrons to form atoms of hydrogen and helium. In the process of forming these atoms, subatomic particles became bunched in a relatively small fraction of the total volume of space, and light, generated by the Big Bang, was then free to travel in straight lines across the expanding universe.

The short wavelengths of this primordial light are believed to have been stretched by the continuing expansion of the universe until the light was transformed into the relatively long wavelengths of microwave radio energy. In 1964 these microwaves were discovered coming from all directions in the sky, a phenomenon known as cosmic background radiation.

Several billion years after the Big Bang, hydrogen and helium atoms began to assemble at various places throughout the universe, and it has been suggested that filamentary patterns shown by superclusters of galaxies may trace this initial condensation. It also has been theorized that quasars, which were first discovered in 1960 through the use of radio and optical telescopes, represent an early stage in the development of individual galaxies, a basic unit in today's universe. Quasars appear to be much smaller than galaxies, yet each seems to produce far more energy than any known galaxy. Quasars are seen towards the outer limits of the observed universe at distances estimated at between 2 and 12 billion light-years from earth. The light of these objects therefore has been traveling through space for between 2 and 12 billion years, and an apparent absence of quasars any closer to earth suggests that their light is a relic from an earlier period in the evolution of the universe.

Galaxies are distributed in groups called clusters and superclusters. The Milky Way Galaxy, where the earth is located, belongs to a small cluster known as the Local Group, which contains just over twenty galaxies, including the Great Galaxy in Andromeda. The band of light seen reaching across the sky on clear nights is an edgewise view of the disk of the Milky Way. This disk contains most of the Galaxy's stars, and also large quantities of dust and gas.

A process related to the revolution of stars, nebulae, and dust around the center of the Milky Way, and the mutual gravitational attraction of these objects, is believed to produce what is known as the Galaxy's system of spiral arms. These arms are regions where stars, nebulae, and dust temporarily become more concentrated as they orbit the Galaxy's center. These arms contain bright young stars and clouds of dust and gas of relatively high density and are also characterized by regions of hot, glowing gas and giant molecular clouds in which new stars are apparently being produced. Such clouds are vast, massive, and consist of accumulations of molecular hydrogen and traces of other molecules, among them carbon monoxide, water, ammonia, and alcohol at temperatures of less than 30°K.

One of the closest giant molecular cloud complexes is located in the direction of the beautiful constellation Orion. Activity in these clouds led to the formation of many bright stars seen in this lovely area of the winter evening sky.

Stars are huge spheres of gas that shine because they are extremely hot. Their high temperatures are the result of thermonuclear fusion reactions deep within the stars, triggered and sustained by the gravitational energy of the star's gas.

How is a new star made? The densest portions of molecular clouds may collapse to form new stars, or outpourings of gas from other stars might trigger the collapse of portions of the clouds, leading to the formation of new stars. As these "protostars" contract, they increase in pressure and temperature and may become visible to observers using infrared telescopes. When the protostar's core temperature reaches about 10 million degrees Kelvin, thermonuclear fusion begins in its core and the object becomes a full-fledged star.

The brightness, temperature, and life expectancy of stars are closely related to the amount of mass that they contain. Stars having low masses compared to the sun shine with relatively feeble light, but their conservative use of hydrogen fuel gives them life expectancies of many hundreds of billions of years. The most massive stars, on the other hand, consume fuel at a furious rate, shine brilliantly, and burn themselves out in the relatively short period of a few million years or less.

During most of a star's life, it produces energy by fusing hydrogen nuclei to form helium nuclei in its core. Stars at this stage are called "main-sequence" stars because they, like 90 percent of the stars in the region near the sun, appear in the central band of the Hertzsprung-Russell (H-R) diagram, a graph that positions stars according to their surface temperature and luminosity. When stars use up the hydrogen fuel in their cores, their energy production shifts to concentric shells surrounding the cores. As this happens, the stars swell to many times their original size, becoming giants. Eventually helium fusion reactions begin in their cores, supplementing the shell's burning of hydrogen. Stars at this stage are customarily known as "red giants" because the largest component of their visible light is from the red portion of the spectrum. However, the blend of all colors produced by these stars appears to our eyes as yellow or yellow-orange.

Some giants for a time may become somewhat unstable and appear as "variable" stars: stars whose magnitude varies periodically, of which the North Star is an example. In time, aging giants loose increasing quantities of gas and dust into space in the form of stellar winds. The majority of stars will eventually expire after losing most of their gas in this way. Their cores are compact enough to avoid dissipation and will survive as white dwarf stars, shining faintly with vestigial energy.

A small percentage of stars, with initial masses greater than about 10 times that of the sun, end their existence in a far more spectacular way: they explode as supernovae. Such an explosion may destroy all traces of the old star's core, leaving only an expanding cloud of gas and dust along with a brilliant flash of light. In other cases, the stellar core may survive and have sufficient mass so that it collapses into a ball of tightly packed neutrons, 1,000 times smaller in diameter than a white dwarf. As these neutron stars spin, some may emit beacons of energy and be observed from earth as pulsars. In some cases, even the super-hardness of closely packed neutrons may not prevent the total collapse of the most massive remnant cores. Cores that survive the supernova explosions of the most massive stars may retain so much mass and gravitational energy that they collapse totally, forming black holes in space.

As old stars fade into oblivion or blow themselves apart, some of their released energy may trigger the birth of a new generation of stars within nearby molecular clouds. These new stars may, in turn, incorporate part of the material that was ejected into space during the last stages in the evolution of their predecessors, stages in which nuclear processes forge the heavy chemical elements found in our bodies and throughout the universe.

2

How to Use This Book

THE Star Guide is a unique system for learning to identify the brightest stars seen from the continental United States.

As the earth rotates and carries us eastward during the night, a progression of stars appears to rise above the eastern horizon. Simultaneously, other stars seem to descend in the western sky until our view of them is blocked by the edge of the earth, and we say that those stars have set. A similar apparent shift in star positions is produced by earth's revolution around the sun. As a result, the evening sky is filled with a different set of stars during each season of the year.

In *The Star Guide,* 105 of the brightest stars in the sky are introduced individually on a date during the year when they are visible in the eastern sky in the evening. They will appear in an imaginary zone we will call the Sky Screen, a narrow band in the sky that is parallel with the horizon and about one-third of the distance from the horizon to the zenith, and that runs from north through east to south. A year-long procession of bright stars is presented on this Sky Screen, as guests would be at the entrance to a great celebration. These featured stars are described as unique individuals, each with a special character and nature. After you've made regular observations for a few weeks using *The Star Guide,* the gradual shift of stars towards the west will begin to fill the sky with familiar stellar friends. These stars will then serve as reference points to guide you in aligning the book's constellation maps, which will then help you explore for additional celestial features.

The following steps will help you learn how to use *The Star Guide.* After you have successfully followed the procedures a few times, many of the steps can be dispensed with. For example, you will soon remember the location of north, and after you have used the "compass card" (page 13) several times at your observing site, you will be able to use landmarks to locate compass directions—a distant tree may indicate east, a certain low building southeast, and so on. You can then find directions by referring to these landmarks and dispense with the compass card. Before long you will be able to visit your observing site with just a flashlight and *The Star Guide* and, at the appropriate time, point out and identify additional bright stars.

1. Gather your materials.

2. Determine the viewing time for your location.

3. Select your observing site and find the Sky Screen.

4. Find north.

5. Turn to Part II and read about the star that is featured nearest the date of your planned observation.

6. Find the appropriate Sky Screen direction.

7. Identify the featured star.

1. Gather your materials:

a. *The Star Guide.*

b. A flashlight with several layers of red cellophane, held in place with a rubber band, wrapped around the lens. This will let you see your materials without spoiling your eyes' sensitivity to starlight.

c. A magnetic compass. The compass will be helpful until you have learned to find the North Star.

2. Determine the viewing time for your location. Refer to the time-zone map found on page 10, which suggests when to use *The Star Guide* in your area. The suggested time will vary from about 8:15 P.M. to 9:45 P.M., depending on your location within a particular time zone. Remember to add one hour to this time whenever daylight saving time is in effect. Observations made within about fifteen minutes of the suggested time will show only slight differences in a star's position.

3. Select your observing site and find the Sky Screen. Before you begin to look for stars, find a convenient place near your home that provides a view of the eastern sky and is as free as possible from the glare of electric lights.

Most of the stars described in *The Star Guide* are bright enough to be seen even from a well-lit street if the sky is clear, so if you live in a city and can't get away from electric lights, don't worry. Do avoid the direct glare of the lights by cupping your hands around your eyes or using a light shield, which you can construct from a shoe box: remove one end of the box, cut eye holes in the other, replace the lid, and look through the box as you would a telescope.

Extend your arms fully towards and level with the eastern horizon, then place one hand on top of the other as you measure three fist heights up from the horizon. This is the approximate altitude of the Sky Screen zone. If your horizon is obscured, measure upwards from eye level, again with arms fully extended. Look around the top of your upper fist to check for possible obstructions, such as buildings or trees. You may wish to select several different vantage points in order to provide clear views of all parts of the Sky Screen from north to south.

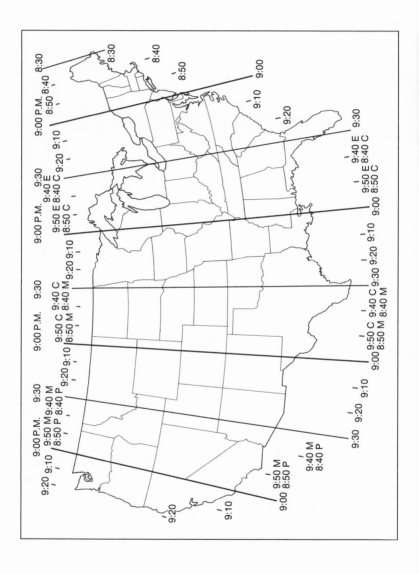

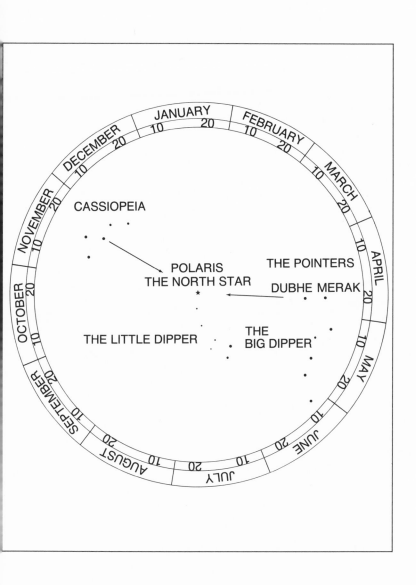

4. Find north. At your observing site, use the magnetic compass to find the general direction of north. In most locations, the magnetic north direction shown on a compass is not exactly the same as the direction of true (geographic) north, our basic reference for finding horizon directions.

Face in the indicated northerly direction and hold the map on page 11 in front of you. Rotate the map so that your current date is positioned at the top of the circle. Refer to this map as you look for the stars of the Big Dipper.

On summer evenings, the Dipper is located about halfway between the horizon and the zenith, in the northwest, to the left of true north. At about 9:00 P.M. on winter nights, the Big Dipper is also located about halfway up in the sky, but in this case it is seen in the northeast, to the right of true north. At your suggested observing time during the spring season, the Dipper is found nearly overhead, close to the zenith.

After you have located the Big Dipper in the sky, trace its figure with your finger and identify Dubhe and Merak, the two "pointer" stars, which are located at the front of the Dipper's bowl. Then, with the help of your map, trace an imaginary line in the sky from Merak past Dubhe until you reach Polaris, the North Star. Notice that the distance from Polaris to Dubhe is about five times greater than the distance from Dubhe to Merak. Also note that Polaris is neither a particularly bright star nor does it have any bright neighbors in its immediate vicinity.

During the autumn season, the Big Dipper is located low in the evening sky, near to the horizon, as may be seen in the photograph on the front cover of *The Star Guide.* When looking for the North Star during autumn, at around 9:00 P.M. you may find it more convenient to refer to the M-shaped pattern of stars in the constellation Cassiopeia. At your suggested observing time during autumn, Cassiopeia is located nearly overhead as you face north. After you have identified stars in Cassiopeia, use the guideline shown on map on page 11 to help locate the North Star.

The point on the horizon beneath Polaris approximates the direction of true, geographic north. After some practice using the above steps, you will be able to turn to the North Star as a quick and reliable direction indicator.

5. Turn to Part II and read about the star that is featured nearest to the date of your planned observation. Study that star's finder map and note the star's Sky Screen direction. Then take your materials and plan to arrive at the observing site shortly before the time suggested for using the Sky Screen maps.

6. Find the appropriate Sky Screen direction. Turn to the "compass" on page 13. Hold the page horizontally. Align the compass's north point with true north on your horizon. Next, look across the

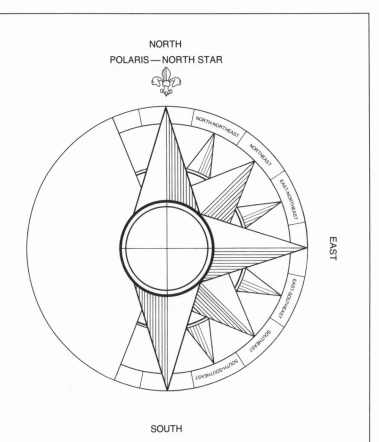

Align the north direction of this card with the point on the horizon directly beneath Polaris. Sight through the center of the compass to find the other horizon directions.

top surface of the page, sighting through the compass center towards the direction indicated for the night's featured star. For example, on April 5, the featured star is Arcturus, which has a Sky Screen–horizon direction of east. Aim your arms in the appropriate direction and measure upwards from the horizon a total of three fist heights. The level of your top fist will point you towards that part of the Sky Screen where the featured star is located.

7. *Identify the featured star.* Note any patterns formed by bright stars in the section of sky surrounding your top fist, then compare the relative positions of these stars with those shown on the featured star's finder map. The featured star itself is located at the map's center, represented by a graphic star. Other bright stars in the map's field of view are indicated by dots. Remember that, with your arm fully extended, your hand is about 10 degrees of arc wide, and if you extend your little finger, it would just cover the disk of the full moon, which has an apparent diameter of about one-half a degree of arc.

The next section tells you more about the information given on each finder map.

How to Use the Star Finder Maps

Each Sky Screen finder map shows the featured star at its center, represented by a star-shaped symbol. These symbols are graded in size according to the apparent magnitudes of the featured stars, as are the dots used to represent other bright stars in a map's field of view.

Stars shown on these finder maps will generally belong to several different constellations, but because they are relatively bright, many of these stars form patterns that are considerably easier to find and remember than are the traditional constellation figures, which often include many rather faint stars. For example, the "bent-line" pattern outlined by Arcturus and its neighbors Muphrid, Izar, and Alphecca is much more obvious than is the traditional outline of the constellation Boötes.

The featured star is usually about three fist heights from the horizon, which is, on the finder map and in the sky, equal to about 30 degrees of arc. (A few stars with considerable southern declinations, such as Adhara in Canis Major and Shaula in Scorpius, never attain an altitude of 30 degrees when observed from midnorthern latitudes. Since the highest point of their paths across the sky is reached at the celestial meridian, we will therefore face directly south when we use the finder maps to identify these stars.)

At the bottom of each Sky Screen finder map you will find the appropriate horizon directions to show you which way to face when you measure three fist heights up from the horizon. You should try to pick out landmarks at your observing site so that you can easily spot "south-southeast" or "north-northeast."

Location of the Sky Screen

Example Arcturus

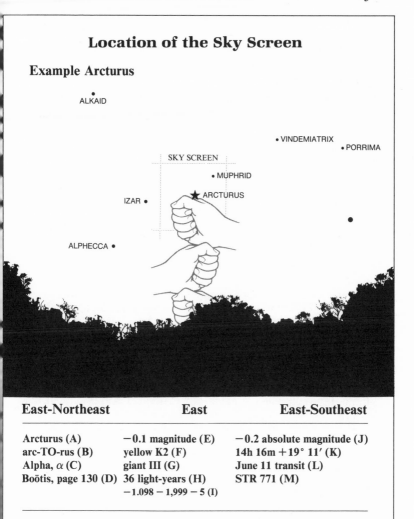

East-Northeast	East	East-Southeast
Arcturus (A)	−0.1 magnitude (E)	−0.2 absolute magnitude (J)
arc-TO-rus (B)	yellow K2 (F)	14h 16m +19° 11′ (K)
Alpha, α (C)	giant III (G)	June 11 transit (L)
Boötis, page 130 (D)	36 light-years (H)	STR 771 (M)
	−1.098 − 1,999 − 5 (I)	

Note that these horizon directions have been rounded off to the nearest compass point, so small variations may be seen.

How to Use the Stars' Data Tables

Each featured star is provided with a data table, which gives information about its basic characteristics. Each lettered element in the table for Arcturus is explained below.

A. PROPER NAME: ARCTURUS

Arcturus is one of the few stars that had a proper name in classical antiquity. Most stars were identified in Ptolemy's *Syntaxis* of the second century A.D. only by coordinates and descriptive positions within mythological constellation figures.

Paul Kunitzsch, a linguistics scholar who has extensively researched the origins of star names, found that most of the names in use today come to us from several sources. Some, such as Aldebaran, are of pre-Islamic Arabic derivation, whereas others are based on Arabic translations of the star position descriptions found in the *Syntaxis.* An additional group of Arabic-sounding star names is actually of European origin. These, Kunitzsch suggests, originated as either intentional or accidental modifications of Arabic words and were adopted as a result of the work of various individuals, including Joseph Scaliger, a French philologist who lived during the end of the seventeenth century, and Giuseppi Piazzi, an Italian astronomer who published a star catalogue in 1814.

B. PRONUNCIATION: arc-TO-rus

C. GREEK LETTER DESIGNATION: ALPHA, α

In 1603, Johann Bayer of Bavaria introduced a systematic method of stellar nomenclature based on letters of the Greek alphabet. Bayer first used this system in his beautifully illustrated star atlas, the *Uranometria.* This work's popularity helped to ensure the wide use of Greek-letter star designations.

The letters were assigned to stars, usually in an order based on a star's relative brightness within its constellation, beginning with *alpha* for the brightest. The full designation also included the genitive form of the constellation name. Since Arcturus is in Boötes, its full designation would be Alpha Boötis, or α Boo.

The constellation descriptions in *The Star Guide* include the genitive forms and abbreviations used with the Greek-letter star nomenclature. You will note that some of the bright stars in this guide are known only by this designation; for example, Zeta Herculis and Eta Draconis. Professional astronomers favor using this and other systematic methods of star nomenclature.

D. CONSTELLATION: BOÖTIS, PAGE 130

Constellation maps and descriptions appear throughout the book. A constellation is described immediately after the first of its stars is introduced. In this case, the map for Boötes appears on page 130, following the finder map for Muphrid, March 31.

Most constellations were devised in ancient times when various patterns of stars were used to represent mythological animals or people. Astrophysicist and astronomical historian Owen Gingerich of Harvard University suggests that the widespread identification by indigenous peoples in Europe, Siberia, and North America of stars in

Ursa Major as a "great bear" may indicate that this constellation originated as long ago as the last Ice Age, when migration from Asia to America was possible across the Bering Strait land bridge.

Clay and stone artifacts suggest that some constellations, such as Leo, Virgo, and Taurus, may have been known in Sumeria before 3000 B.C. There are further indications that many of our present constellations originated in Mesopotamia before the first millennium B.C. Professor Gingerich has also noted that the earliest Babylonian records of zodiacal constellations date to about 700 B.C.

Mesopotamian constellation outlines were introduced into Greece during the early part of the first millennium B.C. and many were given Greek mythological associations by about the fourth century B.C. In the second century A.D., Ptolemy described a total of forty-eight classical constellations.

Ancient pictorial representations of these constellations were not available in medieval Europe, and as a result, Albrecht Dürer in the year 1515 designed a set of constellation figures based on Ptolemy's written descriptions. Dürer's designs in turn inspired those in Bayer's *Uranometria* and thereby form a basis for our modern constellation figures.

In 1930, the International Astronomical Union, the world organization of professional astronomers, formally defined a total of eighty-eight constellations as regions of the celestial sphere having specific boundary lines. These included constellations visible only in the southern hemisphere and several faint northern constellations, all of which were devised only during the past few centuries.

After you have learned to recognize and identify the brightest stars in a part of the sky, turn to the constellation references for help in finding fainter stars and other objects. These fainter objects are also described in the "Nearby Features" sections of the star descriptions.

Objects listed under "Nearby Features" are of several varieties. Some are rather esoteric phenomena recently discovered by astronomers. These usually cannot be observed without sophisticated equipment, yet it is satisfying to know where some of these objects are located. Nebulae and galaxies generally appear as pale wisps of light in a telescope, even under clear, dark skies. Most nebulae and galaxies are not visible unless the observing conditions are so fine that the Band of the Milky Way can clearly be seen with the naked eye.

Star clusters and double stars provide some of the best views for observers with small (100×) telescopes. In the case of many double stars very dark skies are not required for observations. In fact, slightly hazy summer evenings afford the stable atmosphere required to resolve close doubles. On the other hand, cold turbulent winter skies usually yield the clearest views of nebulae and galaxies.

E. APPARENT MAGNITUDE: −0.1

During the second century B.C., the Greek astronomer Hipparchus devised six categories or "magnitudes" to describe the relative brightness of stars. The brightest stars were said to be of the first magnitude, while the faintest visible to the eye were assigned magnitude six.

In 1856, Norman R. Pogson refined the traditional magnitude system. He noted that first-magnitude stars are about 100 times brighter than those of the sixth magnitude and therefore defined an intensity ratio of about 2.5 between each category of magnitude.

The star Vega has long been used to help calibrate the magnitude scale, and its magnitude was defined as 0.00. Photoelectric measurements and comparisons of star brightnesses with that of Vega serve to identify the magnitudes of other stars.

Increasingly higher positive numbers are assigned to fainter stars. With the unaided eye most people are able to see objects up to about magnitude 6.5, if observing conditions are favorable. Negative values are used to indicate brilliant objects, such as Arcturus (-0.1), Sirius (-1.5), and the planet Venus (-4.4 at its brightest), that are brighter than Vega. The full moon has a magnitude of -12.5, and the sun shines with an apparent magnitude of -26.8.

F. COLOR AND SPECTRAL CLASS: YELLOW K 2
The color descriptions given in *The Star Guide* are apparent colors based on those suggested in 1981 by astronomers Howard Cohen and John Oliver as representative of the hues actually perceived by our eyes.

The letters *O, B, A, F, G, K,* and *M* are categories of spectral classification in use today. The designation "p" for "peculiar" is sometimes used to identify a star some of whose spectral lines have a rather unusual appearance. Each of these spectral classes is further divided into ten subunits. Spectral classes are indicative of various stellar characteristics, including surface temperature. The following list gives the approximate surface temperatures for main-sequence stars of various spectral classes.

Apparent color	Spectral class	Temperature (degrees Kelvin)
	O5	40,000
	B0	28,000
	B5	15,500
blue-white	A0	9,900
	A5	8,500
	F0	7,400
white	F5	6,600
	G0	6,000
yellow-white	G5	5,500
	K0	4,900
yellow	K5	4,150
yellow-orange	M0	3,500
	M5	2,800

Astronomers use the Kelvin, or absolute, scale for describing the temperatures of stars. Kelvin degrees are the same size as Celsius degrees; the zero point of the Kelvin scale is equal to -273 degrees Celsius.

Although most stars in the sun's neighborhood of the Milky Way each contain about 70% hydrogen, 28% helium, and 2% other chemical elements, the appearances of their spectra vary considerably. Differences in surface temperature and pressure contribute to the broad array of spectral categories. Temperature and pressure are, in turn, primarily dependent on the initial mass and age of each star.

G. LUMINOSITY CLASS: GIANT III

The following categories of star luminosity were introduced in 1943 by William W. Morgan, Philip C. Kennan, and Edith Kellman of Yerkes Observatory.

Ia	**most luminous supergiants**
Ib	**less luminous supergiants**
II	**bright giants**
III	**normal giants**
IV	**subgiants**
V	**main-sequence stars**

As the core temperature in a contracting protostar reaches about 10 million degrees Kelvin, thermonuclear fusion of hydrogen to helium begins and the contraction ceases. The new star joins what is called the main sequence, a reference to its position on the H-R diagram. A star's location along the main sequence is almost entirely dependent upon its total mass, with the most massive stars located at the top left of the sequence and the least massive stars at the lower right.

As stars use up the supply of hydrogen in their cores, they begin to expand and leave their main-sequence locations on the H-R diagram to become giants. As a result of the increase in diameter, the outer layers of the stars decrease in density and the surface temperatures fall.

The duration of a star's main-sequence stage is primarily dependent on its total mass. The most massive stars evolve into giants after a few million years or less on the main sequence, while a star of average mass, such as the sun, requires about 10 billion years.

Subgiants represent the first stage of evolution away from the main sequence. The majority of stars evolve into normal giants, and the most massive of main-sequence stars will evolve to become supergiants. Such stars "race" through all their evolutionary stages in just a few million years, and as a result, supergiant stars are comparatively young even though they are near the end of their evolution. Some examples of H-R diagrams are found in Appendix C.

H. DISTANCE: 36 LIGHT-YEARS

This information tells us that light from the star Arcturus travels through space for 36 years before it reaches earth. The light-year is a unit of distance (not time) often used in reference to stars. One light-year is the distance through a vacuum that light will travel during the period of one year. A light-year is equal to about 9.5 trillion kilometers, or 63,000 times the distance between the earth and the sun.

Astronomers generally describe interstellar distances in units called parsecs rather than light-years. A parsec is equal to approximately 3.26 light-years, which is the distance from earth at which a star would be seen to have a parallax of 1 arcsecond. Parsecs are more frequently used by astronomers because they can be determined directly from measurements of stellar parallax.

Most of the bright stars visible to the unaided eye are located within about 2,000 light-years of earth. Estimates become considerably less precise with increasing distance and are therefore rounded off to increasingly greater amounts.

I. PROPER MOTION AND RADIAL VELOCITY: $-1.098 - 1.999 - 5$

A star's proper motion is the distance it moves on the celestial sphere each year due to its intrinsic movement through space. This movement is expressed in terms of equatorial coordinates (see below). The first value, -1.098, tells us that Arcturus moves 1.098 arcseconds in right ascension. The minus sign indicates that this is in the direction of decreasing right ascension, or westward. Proper motion in declination is indicated by -1.999. Every year Arcturus moves in the direction of decreasing declination, or southward, by 1.999 arcseconds.

Radial velocity is a motion towards or away from the sun. In the example of Arcturus, the radial velocity is minus (decreasing distance from the sun) at a rate of 5 kilometers per second. If we use the Pythagorean theorem and find the square root of the sum of the squares of a star's proper motion in both right ascension and declination, we obtain the star's *composite* proper motion. Combining the composite proper motion with radial velocity gives us a value for a star's total motion, relative to the Sun. This quantity is known as a star's space velocity.

J. ABSOLUTE MAGNITUDE: -0.2

The absolute magnitude of a star indicates how bright it would appear if it were observed from a standard distance of about 32.6 light-years.

Use of absolute magnitudes provides a way to compare the actual brightnesses of stars without having to consider brightness differences caused by variations in distance from the earth.

Because the actual distance of Arcturus is almost the same as the absolute-magnitude standard distance, the absolute and apparent magnitudes of this star are nearly the same.

K. Equatorial Coordinates:
14h 16m +19° 11'

You do *not* have to use celestial coordinates in order to find bright stars with the help of *The Star Guide*. However, once you have learned to locate and identify the brightest stars, you may wish to use these prominent stars as celestial markers when you search for fainter objects. When you reach this stage, a general knowledge of where bright stars lie on the equatorial-coordinate grid will facilitate the use of many star maps and catalogues. An explanation of these coordinates comes later in this chapter.

L. Transit Date: June 11

The transit date indicates when this featured star crosses the celestial meridian at about 10:00 p.m. daylight saving time (DST). The crossing of the celestial meridian marks the midpoint of a star's east-west passage across the sky, and it also represents the star's highest possible altitude above your horizon. At its meridian transit, the light from a star passes through a thinner section of atmosphere than at any other time, thus the light is usually least affected by atmospheric disturbances.

The transit date provided in *The Star Guide* also marks the middle of a celestial object's evening observing season. Note that Arcturus takes about nine weeks to move from its 9:00 p.m. Sky Screen position to its 9:00 p.m. (10:00 p.m. DST) meridian transit. After another nine-week period, about mid-August, Arcturus will be located in the western sky about 30° above the horizon at 10:00 p.m. DST. During late August, at this time of evening, Arcturus moves progressively lower in the sky as it fades from prominence.

M. STR Numbers: 771

The STR (Star Time Reference) numbers provided in the data table give you a convenient way to extend *The Star Guide*'s system so that you can identify many stars on any night during the year.

The numbers represent the sequence of minutes between the Sky Screen arrival times of featured stars. Refer to Appendix A for an explanation of how to use these numbers in planning your observing schedule.

The Celestial Sphere and Equatorial Coordinates

The celestial sphere is an imaginary surface of indefinite size in the sky that has as its center the observer's location. This sphere provides a model on which we may describe the positions, separations, and motions of stars and other celestial objects.

The point on the celestial sphere located directly above our heads is called the zenith. The astronomical horizon is defined by a great circle 90 degrees of arc from the zenith. Points along the horizon may be identified by various compass directions, such as north, northeast, and north-northeast.

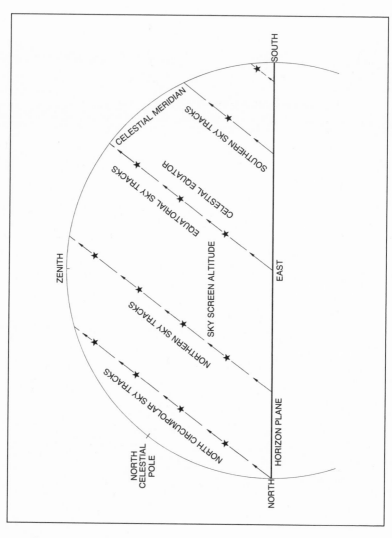

The celestial sphere

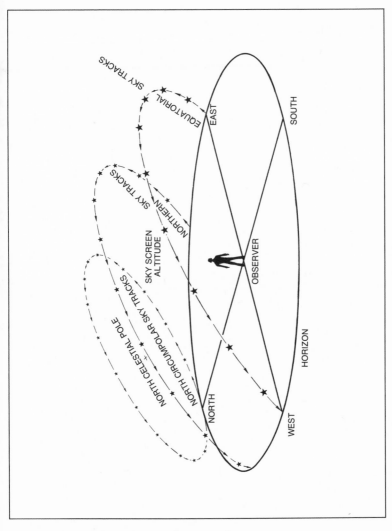

Sky tracks with an observer facing northeast

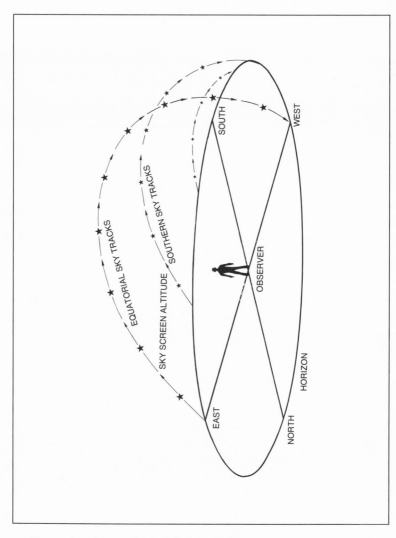

Sky tracks with an observer facing southeast

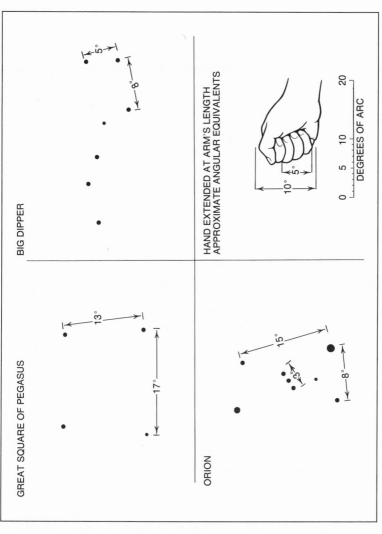

Examples of angular measurements in the sky

A great semicircle that extends across the sky from the south point on the horizon through the zenith and on to the horizon's north point is known as the celestial meridian. This meridian represents the location at which a celestial object such as a star reaches its highest altitude above the horizon, as it intersects or transits the meridian at a 90-degree angle during its apparent motion from east to west.

The celestial poles are at the intersection of the celestial sphere with extensions of the earth's poles. Halfway between the celestial poles lies the great circle of the celestial equator, and each of these features is located directly above the corresponding feature on earth. As a result, if you look up from the earth's equator, the celestial equator passes directly through your zenith, and if you were at either of the earth's geographic poles, the appropriate celestial pole would be seen at the zenith.

Various systems are used by astronomers to identify points on the celestial sphere, but the most generally used is the equatorial coordinate grid of right ascension and declination.

The right-ascension coordinate corresponds somewhat to longitude on a terrestrial map. However, instead of the angular measurement units used to express longitude, right ascension is shown in units of time: hours, minutes, and seconds. A total of 24 hours of right ascension is measured along the celestial equator in an eastward direction beginning at the point known as the vernal equinox, the sun's position at the start of spring, which marks 0 hours of right ascension. The autumnal equinox, 180 degrees of arc along the celestial equator from the vernal equinox, marks the 12-hour coordinate of right ascension. Hour coordinates of right ascension extend north and south from the celestial equator and converge at the celestial poles, forming a pattern similar to that seen with segments of an orange.

Each hour of right ascension is divided into 60 minutes of time, and each minute is divided into 60 seconds of time. The right-ascension coordinate was designed to provide a link between east-west positions of stars and the passage of time on a clock. For example, if the east-west separation of two stars is equal to 1 hour, 32 minutes, and 17 seconds of right ascension, the second star will follow the first across the celestial meridian after a period of 1 hour, 32 minutes, and 17 seconds of time as measured on a sidereal clock.

To help understand differences between sidereal and conventional clocks, consider the following statements: (A) One day (24 hours) is equal to the time period between noon one day and noon the next day. (B) One day (24 hours) is equal to the time period it takes earth to rotate once on its axis. Are these two periods equal? No, they are not, and let us see why. Noon may be defined as the time when the sun transits (crosses) the celestial meridian, and one complete rotation of earth may be defined as the period between successive transits of the meridian by a star other than the sun. These periods are not exactly equal because, as earth moves along its orbit, the sun is seen from a slightly different vantage point each day and as a result it appears to move slowly along the ecliptic towards the east. No such motion is seen for other stars because their distance is so much greater than that of the sun. Because the sun moves relative to the celestial coordinate system, more than one complete rotation of earth is required to bring the sun back to the meridian, and a solar day is therefore about 4 minutes longer than a sidereal day. To account for these 4 extra minutes, conventional clocks tick slightly slower than do sidereal clocks, and during the period a sidereal clock measures 24 hours, a conventional clock registers only about 23 hours and 56 minutes. As a result, relative to conventional clocks, stars appear to rise about 4 minutes earlier each night.

The celestial coordinate of declination directly corresponds to latitude as measured on earth maps. Declination is measured from the celestial equator to the celestial poles in angular units of degrees of arc. Objects in the southern hemisphere of the sky are distinguished with negative declination signs and those in the northern hemisphere by positive signs.

In the following example, the equatorial coordinates of the star Arcturus are given:

Star name	Right ascension (R.A.)	(2000.0)	Declination (dec.)
Arcturus	**14h 15m 40s**		**+19° 10′ 57″**

Star positions given in this book are from *The Bright Star Catalogue,* compiled by Dorrit Hoffleit of Yale University Observatory, with the collaboration of Carlos Jaschek of the Strasbourg Centre de Donnees Stellaires. The number 2000.0 indicates that these are star positions as predicted for beginning of the year A.D. 2000. The last basic reference date was 1950.0, and star coordinates have changed slightly since then due to both intrinsic motions of stars as they travel through space and a shift in the position of coordinate reference lines resulting from the precessional wobble of earth's axis of rotation. Such differences in star positions between 1950 and 2000 cannot be detected by naked-eye observers, but they may be of significance for users of telescopes equipped with setting-circle sighting devices.

Notice the abbreviations *R.A.* used for right ascension and *dec.* for declination. Also note the different way in which minutes and seconds are abbreviated for each coordinate. Although they have similar names, each unit represents a different distance on the celestial sphere.

For convenience, the positions of stars featured in *The Star Guide* are given to the nearest minute of both right ascension and declination. For example, in the case of Arcturus:

Star name	R.A.	dec.
Arcturus	**14h 16m**	**+19° 11′**

Planets

The word *planet* comes to us from ancient Greece, where it meant "wanderer" and was used to describe the motions of Mercury, Venus, Mars, Jupiter, and Saturn as these apparently starlike objects moved about the sky relative to stars in constellations of the zodiac.

In addition to their changes in position, which can often be noticed over a period of days or weeks, we can distinguish stars from planets because the planets shine with a steady light and do not twinkle as do the stars. Therefore, if you discover an additional bright "star" in a zodiacal constellation, and it does not twinkle, it is probably a planet. Mercury and Venus are relatively close to the sun and therefore are never seen in our Sky Screen region during the evening hours. As a further help in identifying planets, consult a current almanac.

3

How We Know about Stars

THROUGHOUT the world, evidence has been found indicating that most ancient peoples made astronomical observations of some kind. Written records from cultures in Mesopotamia, China, India, Arabia, and Greece tell us about systems used to identify individual stars and groups of stars.

Many of the constellations that we use today are believed to have originated in Mesopotamia, and some of these probably date from before the third millennium B.C. By the time of the first millennium B.C., the twelve constellations of the zodiac had been established. Stars in the zodiacal constellations helped ancient astronomers keep track of the positions of the sun, moon, and planets as these objects moved across the heavens relative to the "fixed" stars.

In the second century B.C. the Greek Hipparchus, observing from the island of Rhodes, compiled a catalogue in which about 850 stars were described according to their position in the sky and their apparent brightness, or "magnitude." Many of the astronomical writings of Hipparchus were later included in the *Syntaxis* produced by Claudius Ptolemy in the second century A.D.

Ptolemy was the last great name in ancient Greek astronomy, and his *Syntaxis,* a thirteen-volume treatise on astronomy and mathematics, largely based on the work of others—particularly Hipparchus—has been called the astronomical bible of the Middle Ages. The *Syntaxis* included Hipparchus' star catalogue, descriptions of various astronomical observations and their mathematical analyses, explanations of the construction and use of astrolabes, and also Ptolemy's geocentric (earth-centered) descriptions of planetary motions.

After the ancient civilization of Greece crumbled, the *Syntaxis* survived and in the eighth century A.D. was translated into Arabic by order of Harun al-Rasid, caliph of Baghdad and hero of the *Arabian Nights.* Arabia produced a number of excellent observational astronomers, including Albategnius al-Battani, who compiled tables of solar, lunar, and planetary positions around A.D. 900; and Abul Wafa, who, in the second half of the tenth century, wrote a lengthy book on astronomy and also discovered a subtle variation of the moon's orbit.

In addition to maintaining the ancient astronomical legacy, Arabian astronomers introduced mathematical techniques involving the use of algebra and trigonometry and also, from India, our present system of numbers. Arabia also inspired the majority of the present proper names for stars, and although several, including Sirius and Antares, are still known by names given to them in classical antiquity, most star names in use today have Arabian origins. Arabian astronomy and star nomenclature were brought to Europe by the Moors in Spain, where in the thirteenth century King Alfonso X gathered scholars at Toledo to form a center for astronomical study. This group of astronomers published a detailed catalogue of star and planet positions in the year 1252 known as the Alfonsine Tables. Over the next few centuries, this work was the basic reference for astronomers as well as the vehicle that introduced Arabian star names into European usage. The predictions of planet positions contained in the Alfonsine Tables were based on Ptolemy's geocentric model, and as time passed the predictions became increasingly less accurate.

Ulug-Beg, born in 1394, was a grandson of the Mongol conqueror Tamerlane. He developed an interest in astronomy and constructed a fine observatory in Samarkand. His most important work involved the observation and recording of new positions for the stars listed in Ptolemy's catalogue. His measurements of celestial latitude and longitude were accurate to a few minutes of arc for most stars. These were probably the first good determinations of star positions to have been made since the time of Hipparchus, about 1,600 years earlier.

Modern astronomy began with Nicholas Copernicus, who was born in the Polish town of Thorn in 1473. The contributions made by this great astronomer cannot be overestimated. He broke a 1,400-year tradition of slavish devotion to Ptolemy's geocentric concept and thereby helped to instill a sense of free inquiry in European intellectual circles that strongly contributed to the development of society. He was a contemporary of both Columbus and Luther and with them was a prime mover in the development of the modern world.

Copernicus read the works of some Greek astronomers who suggested that earth both rotated and revolved around the sun. However, these had been rather vague suggestions, without proof, and Copernicus spent most of his later life developing and refining mathematical arguments for a heliocentric (sun-centered) description for the motions of earth and the other planets. His quest for mathematical rigor and, perhaps also, concern about hostile reaction delayed publication of his work, *De Revolutionibus,* until shortly before his death in 1543.

In November of 1572 a brilliant star appeared in the constellation

Cassiopeia where none had been seen before. One of the first to marvel at the new object was a young Danish nobleman named Tycho Brahe. After many careful observations Tycho became convinced that this actually was a new star, and he described it in a book entitled *De Nova Stella.* The report created a sensation because at that time nothing new was ever supposed to appear in the realm of the stars. King Frederick II soon provided funds to build a magnificent observatory for Tycho where he produced star and planet observations many times more accurate than any previously made.

Tycho was one of the most important astronomers in history because of the rich data he left to Johannes Kepler and for the way his observatory, Uraniborg, served as a prototype for great national observatories begun a century later at Paris and London. In addition, Tycho's scientific precision, combined with his flamboyant personal style, captured the popular imagination, thereby presenting astronomy as an expression of human intellect and spirit rather than as an occult art.

Tycho sought to devise a theory for predicting planetary positions that would be even more precise than that of Copernicus. To help analyze his observations, Tycho engaged Johannes Kepler, a young mathematics teacher. Kepler labored for years on the problem, giving particular attention to predicting future positions of the planet Mars. By 1609, Kepler had succeeded and published a report on his planetary studies entitled *The New Astronomy, or Commentaries on the Motions of Mars.* He presented two rules for describing the motion of planets around the sun: first, that the planets travel in elliptical (not circular) orbits, with the sun located at one of the two focus points of the ellipse; and second, that a planet's orbital speed is in inverse proportion to its distance from the sun. In other words, the planets move along their orbits faster when they are closer to the sun. The third of Kepler's laws of planetary motion was introduced in 1619 and is known as the harmonic law. The harmonic law tells us that the time period required for a planet to travel once around its orbit is directly proportional to the size of the orbit.

About the same time that Kepler was solving the mysteries of planetary motion, a Dutch spectacle maker named Hans Lippershey was credited with the invention of the telescope. Early in 1609, Galileo Galilei, a professor of mathematics at Padua, heard about Lippershey's invention. While on a visit to Venice, Galileo constructed one of the new devices and demonstrated it to government officials from atop the campanile of the Cathedral of San Marco. The telescope clearly showed ships several hours before they arrived in port, and Galileo pointed out the instrument's usefulness in navigation and commerce.

By 1610, Galileo had constructed a telescope that magnified objects 30 times, and he used the instrument to discover the four largest satellites of the planet Jupiter and irregularities on the moon's surface, and found that the Milky Way is composed of countless faint stars.

Galileo became a vocal advocate of the Copernican system and in 1632 presented his case in *Dialogue of the Two World Systems.* Although he was later silenced by the Church, his writings and introduction of the telescope to astronomy dramatically accelerated the pace of celestial discoveries.

Until as late as the time of Kepler and Galileo, a widespread interest in astrology provided much of the resources that supported astronomical endeavors. However, as the modern world developed, a new motivation for astronomical study appeared in the form of national commercial development. Nations such as France and England sought an increased share of the world trade, which had been dominated by Portugal and Spain. In order to enhance their maritime positions, these countries sought the advice of professional astronomers to devise an accurate means for a navigator to determine a ship's longitude while at sea, as well as to refine the accuracy of maps.

Construction of the Paris Observatory was begun in 1667 by order of King Louis XIV. In addition to its mandate to develop better means of navigation, the observatory was also a source of considerable royal and national pride. Jean Baptiste Colbert, Louis's finance minister, declared, "The edifice will surpass in beauty and convenience the observatories of England, Denmark, and China, and thus will reflect the magnificence of the Prince who built it" (quoted in *Sky and Telescope,* Vol. 59, February 1980, page 100).

Louis brought a noted Italian astronomer, Giovanni Domenico Cassini, to France to head the new observatory. In 1671, Cassini and his assistant at the new observatory, Jean Richter, obtained a set of observations of the parallax of the planet Mars, which enabled them to determine the distance scale of the solar system.

Parallax is a measurement of the angular difference in position of an object as observed from two different vantage points. Knowledge of this angle, as well as of the distance between observing sites, enables trigonometric calculation of distance to the observed object. For example, Cassini in Paris made observations of the position of Mars relative to adjacent stars at the same time that Richter made similar observations from Cayenne in French Guiana. They could then use the method of trigonometric parallax to calculate the distance to Mars. Knowledge of this distance along with the theories of Copernicus and Kepler enabled calculation of distances to other planets and to the sun.

A few years later, also in Paris, Danish astronomer Ole Rømer was able to measure the speed of light using his observations of the satellites of Jupiter and the knowledge of that planet's distance, which Cassini's work had provided.

Across the English Channel, King Charles II began to recognize the value of astronomy and founded the Royal Observatory at Greenwich in 1675. At Greenwich, the astronomer royal, John Flamsteed, undertook the first mapping of the stars using a telescopic measuring device. His star positions were published under the title *Historia Coelestis Britannica.*

In 1687, Isaac Newton's theories of motion and gravitation were published in a volume titled *Philosophiae Naturalis Principia Mathematica.* Newton's work provided a base for modern physical science and was of enormous importance to astronomy. In addition to his theoretical work, Newton experimented with light and, in 1668, invented an improved type of reflecting telescope, which has been widely used ever since.

In 1718, Edmund Halley, a friend of Newton's who had sponsored publication of the *Principia,* noted that the bright stars Sirius, Procyon, and Arcturus each had changed position in the sky somewhat since the time of Hipparchus. In the case of Procyon, its displacement was more than the apparent diameter of the full moon. Because of the sizes and directions of these position changes, Halley believed that they indicated the stars' actual motions through space rather than being a manifestation of some undetected motion of the earth. He predicted telescopic observations of other stars would in time demonstrate similar motions. Such motions were later measured and are known as the "proper motions" of stars. With Halley's discovery, the ancient concept of fixed stars was abandoned.

Halley is better known for another project, completed in 1705, in which he described the orbits of various comets. He noted that the orbital data for bright comets seen in the years 1531, 1607, and 1682 were so similar that they probably represented observations of the same comet, which Halley predicted would return to the inner solar system about the year 1758. On December 25, 1758, an amateur astronomer named Johann Georg Palitzsch was the first to observe the return of what is now known as Halley's Comet.

One of those who undoubtedly saw the comet was a young musician named William Herschel, who had a considerable interest in mathematics and astronomy. By 1773, Herschel had become director of music for the city of Bath in England, and he worked at a heavy schedule of performing, composing, and teaching. In June of that year, during the summer lull in his responsibilities, Herschel found a chance to cultivate his boyhood affection for the stars. In a matter of months he had become a skilled telescope maker and observer. On March 13, 1781, Herschel detected a strange-looking "star" in the constellation Leo and alerted other astronomers to its unusual appearance. The object turned out to be the planet we call Uranus, the first planet to be discovered beyond the five — Mercury, Venus, Mars, Jupiter, Saturn — known since ancient times.

The discovery of Uranus brought Herschel wide recognition and the support of King George III. He began to devote all of his energy to astronomy and, with the assistance of his sister Caroline, started a systematic program of sky observations. Double stars, star clusters, and nebulae were some of the objects they catalogued. Herschel also counted stars and measured their distribution in various parts of the sky. From this he was able to deduce that the sun is part of a vast, disk-shaped system of stars.

Herschel was probably the first astronomer to study stars extensively as objects of interest in their own right rather than for their use as celestial reference points in navigation and planetary observations. Because of the extent of his research, Herschel is often considered to be the founder of stellar astronomy. He constructed a succession of large telescopes, culminating in his great 40-foot-long reflector. This instrument had a mirror 48 inches in diameter and was by far the largest telescope that had ever been made. Herschel was a scientific pioneer, and the massive barrels of his great reflecting telescopes must have appeared as wonderful incongruities, thrusting above the rooftops of Georgian England.

During the 1780s, another English astronomer, John Goodricke, discovered that certain stars, such as Delta Cephei and Algol, changed slightly in brightness over a period of a few days or less. Along with his friend Edward Pigott, Goodricke specialized in the discovery and observation of variable stars. Goodricke's discovery of variable stars was a major contribution to astronomy. For his work he was elected to membership in the Royal Society of London at the age of twenty-one—an especially remarkable achievement because Goodricke had been totally deaf since childhood.

In 1781, French astronomer Charles Messier made the last of his observations of objects to be included in a catalogue of star clusters and nebulae. Messier had compiled this descriptive catalogue so that he could keep track of objects that might possibly mislead him as he searched for new comets. Although he is credited with having discovered at least fifteen comets, Messier is now best remembered for his catalogue of celestial objects, all of which may be seen with relatively small telescopes. The Messier objects, designated by "M" numbers from 1 through 110, include some of the favorite sights in the sky for telescopic observers.

Messier often used a refracting telescope of the achromatic type invented by Chester Hall in 1733 and developed by John Dollond of London during the 1750s. With the advent of the nineteenth century, the center of refracting-telescope construction shifted to Germany, where Joseph von Fraunhofer produced a series of magnificent instruments.

By the 1830s, astronomer Wilhelm von Struve was using a 9.6-inch-diameter Fraunhofer refractor to make the first extensive survey of double stars. Struve's telescope was located at the Dorpat Observatory in Estonia, and, on its newly designed Fraunhofer mounting, it served as one of the prototypes of the great refractors built in later years. The quality of Fraunhofer's telescope allowed Struve to measure positions of double stars accurately over a period of many years and, using Kepler's Laws and Newtonian mechanics, determine the characteristics of some of their orbits.

In the year 1838, the German astronomer Friedrich Wilhelm Bessel, using a specially designed 6¼-inch Fraunhofer telescope at the Königsberg Observatory in East Prussia, made the observations of stellar parallax required to determine the distance of the star known as 61 Cygni. Bessel was able to detect and measure a minute shift in the star's position between two observations made six months apart from opposite sides of earth's orbit. Using trigonometry, Bessel calculated the star's distance — the first time that the distance to any star other than the sun had been successfully measured. The successful use of this technique has been essential to the development of stellar astronomy.

On February 28, 1843, one of history's most remarkable comets nearly collided with the sun. The object was visible in broad daylight, and as it swept past the sun, large quantities of gas and dust were dislodged and swept back by pressure of the solar wind, forming an immense tail. By the middle of March, observers in the northern hemisphere were able to see the comet's tail after evening twilight as it extended across more than 50 degrees of arc in the southwestern sky.

This spectacular sight astounded observers and encouraged many citizens of Cambridge, Massachusetts, to seek information about the comet from astronomers at the recently founded Harvard College Observatory. Unfortunately, the observatory lacked adequate instruments for providing much information about the great comet. As a result of their disappointment, members of the Cambridge community provided funds for a new facility. Hilltop land was purchased and a 15-inch objective was commissioned in Munich. This telescope equaled in size the largest refracting telescope in the world — the one at the Imperial Russian Observatory at Pulkowa, near St. Petersburg. The great refractor at Harvard was completed in 1847, and it enabled the United States to move towards the front ranks of astronomical science. It also led to widespread popular interest in astronomy and telescopes.

One Massachusetts resident who had a particular interest in the new telescope was a professional portrait painter named Alvan Clark. In addition to being a fine artist and a renowned marksman, Clark had developed skill as an amateur telescope maker. His self-taught knowledge of optics and his superb vision enabled Clark to detect minor flaws in the objective lens of the new Harvard telescope. His ability to identify these flaws combined with the high cost of the instrument inspired Clark to start his own telescope-making company, and in time both of his sons joined the venture. The firm of Alvan Clark and Sons was to become the world's foremost manufacturer of refracting telescopes, and on five successive occasions, they made lenses for the world's largest refractor. The last in this series, the 40-inch refractor at Yerkes Observatory, has never been surpassed.

Another young American of the mid-nineteenth century whose astronomical achievements captured the popular imagination was Maria Mitchel of Nantucket, who on October 1, 1849, made a discovery that changed her life. That night, using a small telescope, she discovered a comet at a position about 5 degrees above the North Star. She received much fame for her discovery, became a professional astronomer, and was the first woman to be elected to the American Academy of Sciences.

As early as 1665, Newton had shown that a glass prism could be used to separate sunlight into all the colors of the rainbow, spread out in an array called a spectrum. By 1802, William Wollaston of England had shown that the solar spectrum contains a number of dark lines seen at right angles to the color distribution. In 1815, Fraunhofer had located nearly 600 of these dark lines in the solar spectrum, and it was later discovered that certain sets of these lines could be produced in the laboratory by passing light through the appropriate gases. By the early 1860s, a number of scientists, including Lewis Rutherfurd in the United States, Anders Ångström of Sweden, William Huggins in England, Pietro Angelo Secchi in Italy, and Hermann Vogel of Germany, were making observations of dark lines in the spectra of stars.

In 1864, William Huggins, a wealthy English amateur astronomer, made a series of observations of stellar spectra by passing starlight through a spectroscope mounted on a telescope built by Alvan Clark, located at his private observatory in the yard of his London home. Huggins observed that certain sets of the dark spectral lines produced by these stars were identical to those produced by the spectral analysis of certain gases in a laboratory. As a result, Huggins concluded that stars contained chemical elements that were also found on earth. As a result of spectral analysis, it became possible to determine the chemical composition of stars.

Huggins also used spectral analysis to show that many of the celestial objects called unresolved nebulae were in fact vast clouds of softly glowing hydrogen rather than distant star clusters. In 1866, he turned his spectroscope towards a nova that appeared in the constellation Corona Borealis. Light from this suddenly bright star passed through Huggins's 8-inch Clark refracting telescope and then through his spectroscope's prisms. Huggins found evidence that a shell of hot hydrogen gas had been ejected by the nova at the time that its brightness suddenly increased.

In 1888, while serving as director of Armagh Observatory in Ireland, the Danish astronomer John Dreyer published a listing of about 15,000 star clusters and cloudlike celestial objects. This "New General Catalogue" was followed by two supplements, the first and second "Index Catalogues." Objects described in these works are usually referred to by their "NGC" or "IC" numbers.

In 1872, Henry Draper had produced the first photograph of a star's spectrum. He used a 28-inch reflecting telescope at his private observatory at Hastings-on-Hudson, New York, to make a spectrogram of the star Vega in the constellation Lyra. Draper later produced the first photograph of the Great Nebula in Orion and did much to demonstrate the power of photography as a tool for astronomical research.

After Draper's death, his widow, Anna, donated his telescopes along with an endowment to Harvard College Observatory for the purpose of initiating a major photographic study of stellar spectra. The project was begun under the direction of Edward C. Pickering in 1886, and by 1889 hundreds of thousands of stellar spectrograms had been obtained. The immense task of classifying all these spectra—accomplished by astronomers Antonia C. Maury, Annie J. Cannon, Williamina P. Fleming, and their colleagues—required decades, but the data provided by the Henry Draper Memorial Catalogue project was essential to the development of astrophysics in the early twentieth century.

During her research in 1877, Maury had noticed that certain stars displayed unusually sharp and well-defined spectral lines compared to those shown by other stars of the same spectral class. Her observations suggested to Danish astronomer Ejnar Hertzsprung that such variations in spectral-line width might indicate significant differences in intrinsic brightness, or absolute magnitude, between stars of the same color.

In order to test his theory, Hertzsprung needed to determine the absolute magnitudes of a substantial set of stars. However, Bessel's technique for measuring distance worked only for stars relatively close to the sun, so Hertzsprung had to rely on an indirect method of estimating distance. For this purpose, in 1905 he compared the colors and proper motions of a large number of stars. Then, on the assumption that distant stars would appear to move more slowly, and have slower proper motions, he was able to estimate the absolute magnitudes of stars in his sample.

After he had categorized the stars in his sample, Hertzsprung discovered that the blue-white stars all seemed to have a high intrinsic brightness, whereas the yellow-orange stars fit into two basic categories, one group being far brighter than the sun and the other group being of far lower intrinsic brightness than the sun. Hertzsprung also confirmed the usefulness of comparing the sharpness of certain spectral lines for the classification of stars according to their intrinsic brightnesses.

In 1914, Henry Norris Russell of Princeton University devised a chart that he used to relate the spectral classes and absolute magnitudes of a sample of stars in the neighborhood of the sun. This type of chart is now known as a Hertzsprung-Russell, or H-R, diagram and it is a powerful tool for illustrating many stellar characteristics.

Just before the turn of the century, Lick Observatory in California had received as a gift a 36-inch reflecting telescope from Edward Crossley, a noted English amateur astronomer. It was a famous instrument and had been used to make some of the earliest photographs of the Orion nebula, but the English weather had limited its use. James Keeler, the director of Lick Observatory, labored to refine the Crossley reflector into a reliable instrument, and he then used it to make a pioneering photographic survey of nebulae and star clusters.

Until the end of the nineteenth century, professional astronomers preferred refracting telescopes, such as those made by the Clarks. The development of the large but often unpredictable reflecting instruments had been largely left to wealthy amateurs such as William Parsons in Ireland, Henry Draper in America, and Andrew Common in England, the first owner of the Crossley telescope. Keeler's success with the Crossley instrument encouraged professional astronomers to develop and rely upon the large reflecting telescopes, which have dominated optical astronomy research during the twentieth century.

George Ellery Hale, a member of a wealthy Chicago family, became as successful at fund-raising to build great observatories as he was innovative in the emerging discipline of astrophysics. After having served as the driving force in the establishment of Yerkes Observatory, which began operations in 1897, Hale's interest in studies of the sun's spectrum induced him to seek the pristine air of California's San Gabriel Mountains as the site for a new observatory. While a student at M.I.T., Hale had invented the spectroheliograph, a device used to study the sun's image in wavelengths of light produced by various chemical elements. In 1904, he installed one of these instruments with a large solar telescope atop Mount Wilson, several miles northeast of Pasadena.

Hale's main colleague in the development of Mount Wilson was George W. Ritchey, who as a boy in Ohio had acquired a love for astronomy from his father and who had begun to make telescopes while still in his teens. Ritchey ground and polished mirrors for both the 60-inch and 100-inch reflecting telescopes that were placed in service at Mount Wilson in 1908 and 1918. For a period of thirty years, the 100-inch Mount Wilson reflector served as the world's largest telescope.

In 1919, British astronomer Arthur S. Eddington organized expeditions to Brazil and West Africa to observe a total eclipse of the sun on May 29 of that year. At the suggestion of Albert Einstein, the expedition photographed stars in the vicinity of the eclipsed sun. Later comparisons of these photographs with others made at night demonstrated that the sun's gravitational field had slightly deflected the path of nearby starlight to an extent that was in close agreement with Einstein's predictions. With this confirmation of Einstein's theory of general relativity, the mysteries of stellar energy production came closer to solution.

During the 1920s, Eddington became increasingly interested in the structure and energy supply of stellar interiors. He realized that if main-sequence stars were actually composed of gases as indicated by their spectra, they would need to have core temperatures of about 15 million degrees Kelvin in order to produce internal pressures capable of resisting gravitational collapse.

Eddington showed in 1924 that main-sequence stars are essentially constant in brightness and size and that their luminosities are directly proportional to their total masses. In 1926, Heinrich Vogt and Henry Norris Russell indicated that the position of a star on the H-R diagram is determined by the object's mass, initial chemical composition, and age. This theorem helps astronomers to calculate the temperatures and luminosities of main-sequence stars having various masses and compositions. Since there is comparatively little difference in the chemical composition of main-sequence stars, the location of a given star on the main sequence is primarily determined by its mass.

This type of research helped to make the H-R diagram an extremely useful tool in astronomy. By providing graphic representation of relationships between the temperatures and luminosities of stars, it was possible to identify sequences that represented stages in the evolution of stars. As a result of its ability to portray variations in the physical conditions of stars, the H-R diagram serves as a basic framework for the study of stellar evolution.

Einstein's theory of special relativity indicated, through the famous formula $E = mc^2$, that it was possible for mass to be converted into energy, and such conversion was recognized by many scientists as the source of stellar energy. Spectral analysis had shown hydrogen and helium to be the most common chemical elements present in stars, and their role in energy production seemed evident.

By 1930, studies of star clusters, such as those carried out by Harlow Shapley at Mount Wilson, had begun to provide a greatly increased pool of stars useful as the subjects for further analysis. Because stars within a particular cluster are all at essentially the same distance from earth, their relative luminosities can be easily compared, and H-R diagram comparisons between members of a star cluster can be readily made.

By the late 1930s, Carl Friedrich von Weizsäcker, Hans Bethe, and Charles Critchfield had outlined the basic nuclear processes by which main-sequence stars derive their energy. By means of the "proton-

proton chain" and the "carbon-nitrogen cycle," hydrogen was transformed into helium and energy. With the presentation of these results, astrophysicists were increasingly able to probe the inner mysteries of stars.

In 1942, M. Schonberg and Subrahmanyan Chandrasekhar of India determined that an evolving star will leave the main sequence of the H-R diagram when it has used up about 12 percent of its initial mass through the conversion of hydrogen to energy. When this happens, the star begins to increase in diameter, and it swells to become a giant or supergiant.

Chandrasekhar had found, several years earlier, that all stars should not be expected to end up as white dwarfs. Through the use of formulas of special relativity and quantum mechanics, he calculated that after having evolved through various giant stages, stars that retained more than 1.4 times the sun's mass would collapse into the form of superdense objects of far smaller diameter than even the white dwarfs. In recent decades, discoveries of objects believed to be neutron stars and black holes appear to provide observational evidence of the remains of stars that exceeded the 1.4 solar-mass limit. For his work in developing a better understanding of the processes of stellar evolution, Chandrasekhar shared in the 1983 Nobel Prize for physics.

Two distinct sets, or "populations," of stars in the Milky Way were discovered by Walter Baade in 1944 as a result of observations made at Mount Wilson Observatory under unusually fine observing conditions: wartime blackouts of the Los Angeles area helped to provide dark skies, free from electric lights. His study of the Andromeda galaxy indicated that stars in the arms of spiral galaxies, such as the sun, are rather different from those seen in the cores and globular clusters of these galaxies. The spiral-arm stars, which Baade called population I, represent objects of young, intermediate, and old ages, each containing small but nevertheless significant amounts of chemical elements other than hydrogen and helium. Baade's population II stars, on the other hand, are ancient remnants from the beginning of the Galaxy's existence, before its disk and spiral-arm system evolved. When these first stars formed, probably more than 10 billion years ago, the heavy elements had not yet been produced. As a result, population II stars consist almost entirely of helium and hydrogen, the only elements in the primordial universe.

Today, these two groups indicate a division of stars according to their location in a galaxy, their age, and their chemical composition. Intermediate populations have also been defined.

The conclusion of World War II brought the resources to complete the long-delayed 200-inch-telescope project of the California Institute of Technology. In 1948, the 200-inch (5-meter) Hale reflector went into operation at Palomar Observatory, 160 kilometers south-west of Los Angeles at an altitude of 1,700 meters. For a generation, this mighty instrument served as the world's largest telescope. It was the crowning glory in George Ellery Hale's career as an observatory builder.

One of the many specialists who contributed to the Palomar project was Russell W. Porter, a former Arctic explorer, artist, and instrument maker. Porter's superb drawings of the proposed tele-scope stimulated and sustained public interest in the project during the long years of depression and war. In addition to his work on the Palomar project, Porter is revered as a mentor of amateur telescope makers in the United States. His enthusiastic efforts in organizing and instructing a band of novice telescope makers in Springfield, Vermont, were described by *Scientific American* in a series of articles beginning in 1925. These stories about the Springfield Telescope Makers and their hilltop clubhouse, "Stellafane," did much to stimu-late similar activity in other parts of the country. The following lines from *Russell W. Porter* by Berton Willard suggest the spirit of discovery that Porter inspired in stargazers.

> Eager to share the excitement he had experienced as an ex-plorer, Porter took his group out for observing sessions. Some-times they met in backyards, and sometimes they took week-end camping trips to find dark sky and open horizons. Imagine their enthusiasm, setting out with instruments powerful enough to explore the universe, made with their own hands, to share the wonder and awe of the early astronomers. They ex-plored the wastelands of the universe as their instructor had explored the wastelands of the Arctic.

Feelings such as these often brighten the hearts of those who study the stars, whether by naked eye from one's backyard or with the powerful tools of astrophysics.

During 1946, Bart J. Bok identified various small dark spots in the Milky Way, which he called globules, as possible sites of new star formation. In the next few decades, instruments such as radio and infrared telescopes would be developed to enable astronomers to probe the mysteries of stellar birth, heretofore hidden by clouds of interstellar dust.

By 1952, Ernst J. Opik and Edwin E. Salpeter had begun to de-scribe processes of energy production in stars that are evolving. They showed that after a star has expanded to become a giant, a new process of nuclear fusion begins by which helium is converted into carbon and energy in the star's core. This helium-burning process coincides with the star's arrival at the red giant stage (red giants actually appear yellow or yellow-orange to our eyes). In the same year, Allan R. Sandage and Martin Schwarzschild devised a formula to estimate the time a particular star spends on the main sequence — a period that represents most of its lifetime.

During 1957, Margaret Burbidge, Geoffrey Burbidge, Fred Hoyle, and William Fowler provided an extensive explanation of the nuclear-fusion processes within giant stars and supernovae, which produce most of the chemical elements in the universe. For his contributions to this effort, Fowler shared in the 1983 Nobel Prize for physics.

The information contained in starlight was being extracted by means of increasingly sophisticated techniques during the 1950s. Astrophysicists including Harold L. Johnson, William W. Morgan, and Bengt G. Strömgren were developing methods for comparing the intensity of light in selected portions of a star's spectrum, thereby revealing subtle physical characteristics. Such methods, which are known as quantitative photometric classifications, are of crucial importance in providing data for the development of theories concerning stellar evolution.

In the 1960s, there began a period that some have already called the Golden Age of Astronomy. Innovations in electronics, computer science, spaceflight, and the construction of large new observatories resulted in an unprecedented series of discoveries. High-speed computers facilitated the study of internal stellar processes, and it became increasingly possible to design models of stars at various stages in their evolution. In 1962, Chushiro Hayashi of Japan used computer analysis to help describe H-R diagram paths taken by protostars as they condense from dust and gas before joining the main sequence. By 1967, the study of stellar evolution had progressed to the point where Icko Iben of M.I.T. was able to describe extensively most of the stages in a star's life. During subsequent years, many researchers, including Alastair G. W. Cameron, Robert V. Wagoner, and W. David Arnett, continued to refine the knowledge of what happens within stars.

During the past fifty years, the techniques of radio astronomy have opened an entire new realm of the universe to our observation and have been a central factor in astronomy's current "golden age." Many new and exotic objects and phenomena have been discovered by means of radio astronomy. The first to detect radio signals originating in space was Karl G. Jansky of the Bell Telephone Laboratories, who discovered signals coming from the Milky Way in 1931. Between 1936 and 1944, Grote Reber, an amateur astronomer and electronics engineer in Illinois, developed improved radio telescopes and completed the first extensive radio survey of the sky. Reber worked alone in this field until after World War II, when radio astronomy gradually began to develop as an important research tool.

In 1954, a huge, 200-foot radio-telescope dish was built at Jodrell Bank in England, and an increasing number of discoveries of radio wavelengths began to be made. During 1960, astronomers discovered the curious objects called quasi-stellar radio sources, or quasars. In 1967, Antony Hewish and Jocelyn Bell used a radio telescope of Cambridge University to discover a startling object that produced rapidly pulsating radio signals. More of these objects, now known as pulsars, were soon identified, and they are believed to be fast-spinning remnants produced as the cores of massive stars collapse after supernovae explosions.

During 1970, researchers at the National Radio Astronomy Observatory at Green Bank, West Virginia, first detected radio signals from a visible star, the supergiant Betelgeuse, which is believed to be a candidate for eventual explosion as a supernova.

Another highlight in radio astronomy was the discovery of numerous types of molecules, such as water, ammonia, and formaldehyde, in space. In 1970, Robert W. Wilson of Bell Laboratories detected radio signals characteristic of molecules of interstellar carbon monoxide. In 1978, these signals were used by Patrick Thaddeus of the Goddard Institute for Space Science and colleagues at Columbia University as a tracer for determining the outlines of giant molecular cloud complexes discovered in the region of the constellation Orion. Using Columbia's wide-angle 1.2-meter radio telescope, Thaddeus and members of his team observed and identified these complexes as potential sites for star formation such as that seen in the adjoining Great Nebula in Orion, which sits like a hot, glowing button on the front surface of the molecular clouds.

In 1980, the Very Large Array of radio telescopes was dedicated at the Plains of San Augustin, New Mexico. This huge system consists of a set of twenty-seven, 25-meter-diameter radio telescopes each able to move on tracks that extend for a total of 36 kilometers in a Y-shaped pattern. As a result of the integration of signals received by the component telescopes, the VLA is able to produce radio images of celestial objects comparable in detail to conventional photographs. The VLA has been used in many areas of research, including studies of regions of star formation and supernova remnants.

Developments in electronics have resulted in many significant astronomical advances, such as the remote control of telescopes and electronic image enhancement. One especially important innovation has been the development of charge-coupled devices (CCD's), which can serve as substitutes for photographic film. CCD's are many times more sensitive than film, and they are increasingly being used to enhance the power of telescopes in recording the light from faint objects.

New electronic techniques also made it possible to construct telescopes able to detect and record information about celestial X-ray sources. In 1970, X-ray sensors on the satellite *Uhuru* were used to produce the first extended survey of X-ray-generating objects in space. One of these sources, now known as Cygnus X-1, was the first potential black hole to be identified. The High Energy Astronomical Observatory-2 satellite, later called the Einstein Observatory, launched in 1978, contained specially designed telescopes able to produce X-ray images of celestial sources such as supernova remnants and quasars. The Einstein project, under the direction of Riccardo Giacconi, opened a new "window" in the exploration of space.

Another such window that has recently been opened is in the infrared, or radiant heat, portion of the electromagnetic spectrum. Developments in solid-state electronics made it possible to construct infrared detectors of high sensitivity. In 1975, a C-141 jet aircraft equipped with an infrared telescope went into operation as the Kuiper Airborne Infrared Observatory. It was from this observatory that astronomers discovered a cluster of infant stars located just behind the Orion nebula and hidden from the view of optical telescopes by clouds of interstellar dust.

By the early 1980s, a number of large infrared telescopes had been constructed on the ground for the purpose of studying phenomena such as dust clouds surrounding supergiant stars and regions of star formation. In an effort to avoid atmospheric moisture and dust, which especially hinders infrared telescopes, the Infrared Astronomical Satellite *(IRAS)* was launched in 1983. This cooperative effort between the United States, the United Kingdom, and the Netherlands soon produced spectacular results. By November of that year, *IRAS* had discovered five comets, including one that came closer to earth than any comet had in over 200 years. In addition, complex features of the dust clouds of the Milky Way were recorded, and in the area of stellar research, a system of solid particles was discovered to be in orbit around the star Vega.

Developments in satellite astronomy were not the only areas of advancement in instrumentation. Of the world's current ten largest telescopes, only one — the Hale Telescope on Mount Palomar — was in operation before the year 1973. During the 1970s and early '80s, a series of powerful new reflecting telescopes were constructed. The first of these new giants, a 3.8-meter Mayall Telescope, was dedicated in 1973 at the Kitt Peak National Observatory, about 80 kilometers southwest of Tucson, Arizona. This large observatory is managed by a consortium called the Association of Universities for Research in Astronomy (AURA) for the National Science Foundation. Among other facilities under AURA's jurisdiction are the Hubble Space Telescope Institute in Baltimore and the Cerro Tololo Inter-American Observatory in Chile, one of several new facilities built in the southern hemisphere, marking the first time that extreme southern portions of the sky may be observed with telescopes of the first rank in size.

Also in 1973, the United Kingdom Schmidt Telescope, a large, wide-angle sky camera, was installed at the Siding Spring Mountain Observatory in Australia. Another large southern hemisphere telescope was added when the 3.9-meter Anglo-Australian Telescope was dedicated at this observatory in 1975.

During 1976, three additional great telescopes were completed. They were the 4.0-meter instrument at Cerro Tololo, the 3.6-meter at the European Southern Observatory at Cerro la Silla, also in Chile, and the huge, 6.0-meter reflector on Mount Pastukhov in the Soviet Union.

The year 1979 saw the completion of four more major instruments. These included the 3.6-meter Canada-France-Hawaii Telescope, the 3.8-meter United Kingdom Infrared Telescope, and the 3.0-meter NASA Infrared Telescope, all built atop Mauna Kea, an extinct volcano in Hawaii. At an altitude of 4,200 meters, this site features what are certainly among the best observing conditions on earth.

In that same year, the novel Multiple Mirror Telescope went into service at Mount Hopkins in Arizona. Operated by the University of Arizona and the Smithsonian Astrophysical Observatory, the MMT features an array of six individual mirrors whose light is combined at a common focus with a total aperture equal to that of a 4.5-meter disk. The multiple-mirror design is being considered for the mighty 15.0-meter National New Technology Telescope, which is now in the planning stage.

Another new international facility is the La Palma Observatory operated by the United Kingdom, Denmark, and Sweden in the Canary Islands. The 2.4-meter Isaac Newton Telescope was moved there from Britain, and during 1984 it was joined by the new 4.2-meter William Herschel Telescope.

The rapid growth in astronomical knowledge has stimulated an unprecedented interest in telescope use, which even these new instruments cannot satisfy. Unfortunately, many observatories, including Lick, Mount Wilson, and Mount Palomar, have been rendered less effective by ever-increasing "light pollution" from surrounding communities. This degradation of seeing conditions contributes to the pressures on existing instruments.

A turning point in astronomical history is expected in 1986 with the launching of the Hubble Space Telescope. Although many ground-based instruments are larger, the Space Telescope has the finest optical system ever made, and it will operate in the pristine environment of space.

The Space Telescope is expected to produce the best images ever seen of celestial objects and, in the process, will be able to see about seven times farther into space than any telescope now in existence. As a result, the Space Telescope will help increase the volume of our observed universe by a factor of over 300 times. In other words, if we represent the presently known universe of stars, galaxies, and quasars by a volume the size of a Ping-Pong ball, the universe visible to the Space Telescope will have the relative size of a basketball.

Plans are also moving forward in the development of a 10.0-meter telescope to be built by the University of California and a 15.0 "New Technology Telescope" that will be constructed by the National Science Foundation.

Plans are also moving forward in the construction of the 10.0-meter Keck telescope being built by the University of California and Cal Tech. Both this instrument and the National New Technology telescope, to be built by the National Science Foundation, will be located on Mauna Kea.

Part II
The Star Guide

List of Star Dates

Date	Name	Constellation	Sky Screen Direction	Page
AUGUST				
1	Sargas	Scorpius, the Scorpion	south	232
2	Kappa Scorpii	Scorpius, the Scorpion	south	234
3	Enif	Pegasus, the Flying Horse	east	236
7	Caph	Cassiopeia, the Queen	northeast	240
11	Scheat	Pegasus, the Flying Horse	east-northeast	244
12	Media	Sagittarius, the Archer	south	246
13	Kaus Australis	Sagittarius, the Archer	south	250
14	Kaus Borealis	Sagittarius, the Archer	south	252
17	Schedar	Cassiopeia, the Queen	northeast	254
18	Gamma Cassiopeiae	Cassiopeia, the Queen	northeast	256
20	Markab	Pegasus, the Flying Horse	east	258
21	Nunki	Sagittarius, the Archer	south	262
23	Ascella	Sagittarius, the Archer	south	264
25	Ruchbah	Cassiopeia, the Queen	northeast	266
26	Alpheratz	Andromeda, the Chained Lady	east-northeast	268
SEPTEMBER				
3	Deneb Algedi	Capricornus, the Sea Goat	southeast	272
6	Algenib	Pegasus, the Flying Horse	east	276
7	Mirach	Andromeda, the Chained Lady	east-northeast	278
17	Almach	Andromeda, the Chained Lady	east-northeast	280
28	Sheratan	Aries, the Ram	east	282
29	Hamal	Aries, the Ram	east	288

List of Constellations

You will find constellation maps and descriptions in *The Star Guide* on the pages immediately following the introduction of the constellation's first bright star to reach the Sky Screen location. For example, the constellation Leo is described just after the description of the star Regulus, which reaches its Sky Screen position on February 3.

The list below indicates when the first bright star in each constellation is featured; the page references then show where the constellation descriptions can be found.

KEY FOR MAPS

Dots as they appear on Seasonal Maps

●	●	•	·	·	·
−1	0	1	2	3	4

apparent magnitude

Dots as they appear on all other maps

●	●	●	•	·	·	·
−1	0	1	2	3	4	5

apparent magnitude

Other objects that appear on Constellation Maps

open star cluster	globular star cluster	diffuse nebula	galaxy	faint, yet noteworthy, object

Letters of the Greek Alphabet

α	Alpha	ι	Iota	ρ	Rho
β	Beta	κ	Kappa	σ	Sigma
γ	Gamma	λ	Lambda	τ	Tau
δ	Delta	μ	Mu	υ	Upsilon
ε	Epsilon	ν	Nu	φ	Phi
ζ	Zeta	ξ	Xi	χ	Chi
η	Eta	ο	Omicron	ψ	Psi
θ	Theta	π	Pi	ω	Omega

January Stars

Polaris   January 1

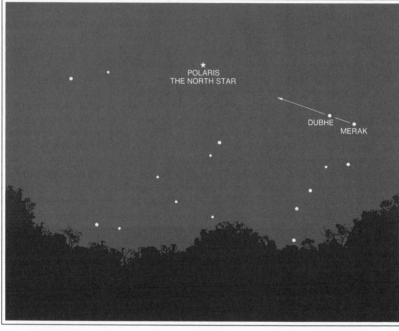

| North-Northwest | North | North-Northeast |

Polaris
poe-LAH-ris
Alpha, α
Ursae Minoris,
    page 62
	2.0 magnitude
	white F7
	supergiant Ib-II
	350 light-years
	$+0.046 - 0.004 - 17$
		-2.9 absolute
		  magnitude
		2h 32m $+89°$ 16′
		December 14 transit
		STR does not apply
		  due to minimal
		  position changes
		  for this star

Location • Polaris is the most famous star in the night sky because it may be found in the same northern direction at all times throughout the year. Also known as the North Star, Polaris has an altitude above the horizon that is about equal to an observer's latitude. Both of these characteristics have given Polaris a special role for travelers and sky observers. Because the star's apparent position varies only slightly, this map and directions are valid throughout the year.

Meaning•The name Polaris is of Latin origin, from Stella Polaris, meaning "the pole star."

Lore•Although it is the best-known star, Polaris is far from being the brightest. About fifty other stars shine more vividly in the sky. Another Latin title was Navigatoria, "the star of navigation." In China, the North Star was designated as the Great Imperial Ruler of Heaven. The term Pivot of the Planets was used as a description in India. This title results from the apparent motion of stars and planets around the North Celestial Pole. Such motion is due to the daily rotation of earth around its axis. Since the axis is pointed in the direction of the celestial poles, celestial objects appear to circle the poles.

During the late sixteenth century, both William Shakespeare and Edmund Spenser called Polaris the Lodestar.

The term Cynosura, meaning "center of attraction," was frequently used in the seventeenth and eighteenth centuries.

History•While constructing his star catalogue in the second century B.C., Hipparchus compared his own observations of star positions with some made several generations earlier. He noticed a slight shift in star positions relative to the North Celestial Pole. He realized that these variations were due to a gradual change in the direction of earth's axis of rotation, which defines the position of the celestial poles.

Astronomers have learned that this shift in star positions relative to the poles is caused by gravitational forces between the earth and the sun, moon, and other planets. These forces cause the earth's axis to wobble like a toy top that is slowing down. As a result, the axis points to a progression of points on the celestial sphere during a cycle of about 26,000 years. The celestial pole represents a standard of position reference. When it moves, the stars apparently move relative to coordinate systems that are linked to the polar direction. The effect that causes this change in the position of the celestial poles is known as precession.

The pole traces out a path in the northern sky with an approximate diameter of 47 degrees during the 26,000-year period. The effect of planetary attraction slightly changes the polar tilt of the earth so that the precessional track of the pole around the sky is not an exact circle. As a result, the pole does not return to precisely the same point in the sky after 26,000 years. Another result of precession is a gradual progression of different pole stars.

For several thousand years, during the time of ancient civilizations, the star Thuban, in the constellation Draco, was considered to be the North Star. Gradually, precession caused the North Celestial Pole to move away from Thuban and to approach two stars at the bowl of the Little Dipper. During the period between about 1500 B.C. and A.D. 500, these stars, Pherkab and Kochab, drew the attention of travelers towards the pole. They came to be known as the Guardians of the Pole and served as twin pole stars until the last days of the Western Roman Empire.

During the last 1,500 years, the North Celestial Pole has been moving closer towards the star Alpha Ursa Minoris, which we know as Polaris, our North Star. By the year A.D. 2102, the Polaris-pole distance will be at a minimum. The separation then will be slightly less than the diameter of the full moon. Thereafter, the distance will increase. The next bright star in the polar succession, Alderamin, will be in the vicinity of the pole about the year A.D. 7000. Subsequent pole stars include Deneb, at about A.D. 10,000, and Vega in A.D. 14,000

Description • In addition to its significance as the pole star, Polaris has other interesting characteristics. During the year 1780, English astronomer William Herschel discovered that Polaris is a double star. The companion has a magnitude of 8.8 and is located about 18 arcseconds from Polaris. This is about equal to the apparent diameter of a dime as seen from the length of two football fields set end to end. We find that Polaris, with an average magnitude of 2.02, appears to be about 650 times brighter than its companion.

A telescope having an objective diameter of 3 inches or more, using a magnification of about 75 power, should be able to show the separation between Polaris and the companion star. The telescope is then said to have "resolved" the pair. The considerable difference in brightness between these two stars makes it rather difficult to see the companion due to the glare of Polaris. Therefore, you may have to examine the telescopic field of view for a while before the secondary star becomes apparent. Polaris and its ninth-magnitude companion represent an example of a double star whose separation is visible in a telescope. Such pairs are called visual doubles. This distinction sets them apart from stars that are revealed to be double by other methods, such as by spectral studies.

Polaris and the companion have the same motion through space and therefore probably form a gravitationally bound system with both stars orbiting around a common center of gravity.

Polaris is a white spectral class F7 supergiant, with a surface temperature of about 6,200°K. This star's spectrum shows prominent dark lines representing ionized metals such as calcium, iron, and chromium. Lines of hydrogen are less apparent than they are in hotter stars, up to about 11,000°K in temperature. Lines of neutral iron, chromium, and other metals are also present. Polaris and other supergiants of class F7 display broader spectral lines than do ordinary giants or main-sequence stars of this spectral classification.

The North Star is also a variable star. Its brightness changes slightly between magnitudes 1.92 and 2.02 during a period of just under four days. Polaris is a member of an important group of such stars known as the Cepheid variables. Cepheids are giant stars that pulsate due to internal energy imbalances. The light variations of Cepheids serve as an important tool for measuring distances in the Milky Way Galaxy and beyond.

The intrinsic brightness or luminosity of Polaris is about 1,600 times that of the sun. The distance to the North Star is estimated to be about 350 light-years. This means that, as we watch Polaris, the light that enters our eyes is completing a journey that began about 350 years ago.

Studies of the spectrum of Polaris indicate the presence of another, invisible companion star. No telescope has shown visually the separation between these two stars. The use of spectral analysis, however, reveals clues to the existence of this close companion.

Ursa Minor, *the Little Bear*
(The Little Dipper)

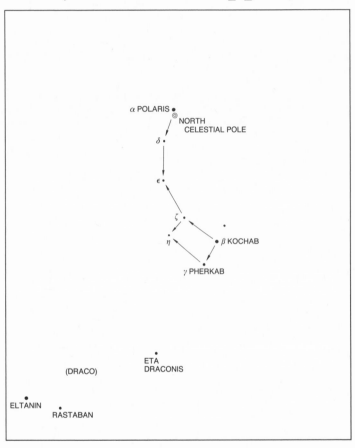

BRIGHT STARS OF URSA MINOR

| January 1 | Polaris | 2.0 | white | supergiant |
| February 27 | Kochab | 2.1 | yellow | giant |

URR-sah MY-ner
U Mi
Ursae Minoris

Evening Season • Ursa Minor is located within 20 degrees of the North Celestial Pole and is above the horizon throughout the year. However, the constellation contains only three stars brighter than fourth magnitude, and clear, dark skies are required for distinguishing the Little Dipper.

Sky Track•North circumpolar

Lore•The name Little Bear was probably introduced into Greece about 600 B.C., from the Middle East. An earlier Greek term for these stars was Cynosura, a name that eventually was applied to the north star of our era, Polaris.

In Greek mythology, the Little Bear represented Arcas, the son of Princess Callisto of Arcadia, who was transformed into the Great Bear when placed in the sky by Zeus. In Arabia, these stars were known variously as the Lesser Bear, the Fish, the Bearing of the Earth's Axis, and as a part of the celestial Fold or Pen. To the ancient Egyptians, the stars of Ursa Minor seem to have represented a Jackal. Vikings recognized these stars as the Small Chariot, or the Throne of Thor. More recent Scandinavians popularly call Ursa Minor the Little Wagon.

In the United States, the name Little Dipper is applied to Ursa Minor's brighter stars.

Description•Polaris marks the end of the Little Dipper's handle. The stars Kochab and Pherkab are located at the front of the Dipper's bowl. The Little Dipper's handle curves away from Polaris towards the handle of the Big Dipper.

Johann Bayer's *Uranometria* shows Polaris at the tip of the Little Bear's tail. The Little Dipper's bowl forms the Bear's body, with its head between Kochab in Ursa Minor and Thuban, a star in the constellation Draco. The Bear's legs are portrayed to the south on both sides of the Dipper's bowl.

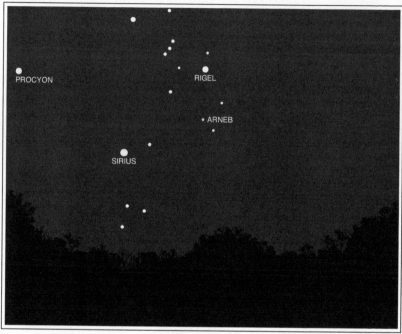

Southeast   **South-Southeast**   **South**

Arneb	2.6 magnitude	−4.5 absolute
ARE-neb	blue-white F0	magnitude
Alpha, α	supergiant Ib	5h 33m −17° 49′
Leporis, page 65	850 light-years	January 29 transit
	−0.006 +0.001 +24	STR 411

Location•Third-magnitude Arneb is found about 10 degrees south of the stars Saiph and Rigel in Orion.

Meaning•The name Arneb is taken from Ulug-Beg's description of the Throne of Orion, formed by the stars Alpha, Beta, Delta, and Gamma, in the constellation Lepus, the Hare.

Lore•Arneb and other stars in Lepus are rather inconspicuous compared with many in neighboring constellations. The curving lines of faint stars may have evoked the association with a hare frantically churning its legs to escape pursuit.

In ancient China, Arneb formed part of an asterism called the Star Screen. Asterisms are patterns of stars that may extend beyond constellation boundaries.

History•In 1835, the English astronomer John Herschel discovered an eleventh-magnitude companion to Arneb at a distance of 35.5 arcseconds. The separation has not changed since that time, indicating a coincidental lineup of these stars on the celestial sphere. Such pairs are known as optical doubles. They are not physical companions in a gravity-bound orbital system as are the components of Polaris.

Description•Arneb is a blue-white supergiant star with a luminosity 5,200 times that of the sun. Its surface temperature is about 7,300°K, and its distance is estimated to be 850 light-years. The spectrum of Arneb is characterized by absorption lines of hydrogen and ionized metals. The hydrogen lines are broader than those of giants and main-sequence stars of class F0. The reverse is true for the ionized metals such as iron and titanium. Arneb moves westward 0.006 arcsecond and northward 0.001 arcsecond each year. Its distance from earth increases by 24 kilometers every second.

Nearby Feature

QUANTRANTID METEOR SHOWER

About the night of January 3, one of the year's major meteor showers reaches its maximum intensity. Under clear, dark skies an observer can usually expect to see more than forty meteors per hour when this shower takes place.

Quantrantid meteors have an average speed of about 40 kilometers per second and appear to streak across the sky from their radiant point between the Big Dipper's handle and the head of the constellation Draco.

Lepus, *the Hare*

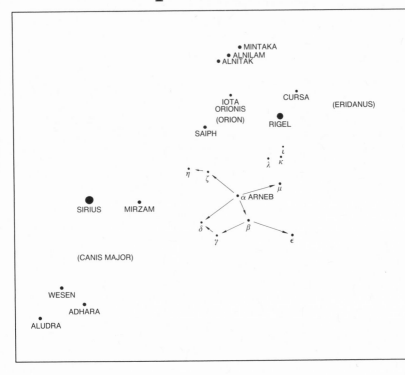

BRIGHT STAR OF LEPUS

January 3	Arneb	2.6	blue-white	supergiant

LEE-puss
Lep
Leporis

Evening Season•January – February

Sky Track•The Hare traverses low in the sky from southeast to southwest.

Lore•Romans described the celestial Hare as "swift," "light-footed," and "eared." These associations could have been inspired by the curving lines of faint stars in Lepus, which originate at Arneb. The lines may have evoked images of the whirling tangle of ears and legs presented by a hare in flight. The proximity of Orion, the Hunter, and his dog, Canis Major, may have contributed to the association of these stars with a hare.

Description•Lepus occupies the section of sky just to the south of Orion and west of Canis Major. The lines of stars in this constellation are most easily seen with the help of binoculars, unless the sky is quite clear and free from the glow of artificial lights. Binoculars provide a fine way to explore the heavens after you have begun to learn the names and locations of the brightest stars. For example, using the Lepus constellation map and binoculars you can try a technique called "star-hopping." Begin at Arneb and look next for Mu, then go to Lambda, Kappa, and Iota. In a similar way, start from Arneb and look for Zeta, then Eta, and Theta. To the south of Arneb, binoculars can help you trace the stars from Delta to Gamma, then to Beta and beyond. Star-hopping provides a bridge from the bright stars you will learn to identify using *The Star Guide* to fainter objects shown on more detailed star atlases.

Procyon   January 5

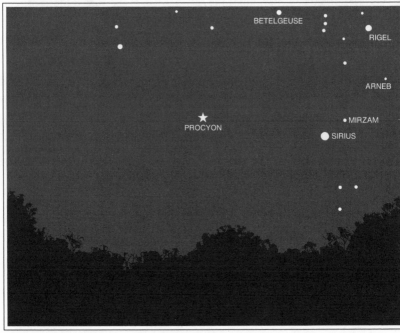

East	East-Southeast	Southeast

Procyon PRO-see-on Alpha, α Canis Minoris, page 70	0.4 magnitude white F5 subgiant – main sequence IV – V 11.2 light-years −0.706 −1.029 −3	2.7 absolute magnitude 7h 39m + 5° 14′ March 2 transit STR 416

Location•Procyon is a very bright star located about 25 degrees east of Betelgeuse. It is nearly the same distance northeast of Sirius. Procyon is located 12 degrees northeast of the centerline of the Milky Way. When the night is clear and dark, the Milky Way appears as a softly glowing band of light reaching across the sky.

Meaning•The name Procyon means "before the dog." This indicates that it rises before, and heralds the arrival of, Sirius, the Dog Star.

Lore•Procyon is the eighth-brightest star in the night sky. It is one of the few stars that has a proper name dating to ancient times. The Chinese called Procyon the Southern River, referring to its location at the edge of the Milky Way Band, which is located just to the southwest of this star. Today, Procyon is sometimes called the Little Dog Star.

History·In 1718, Edmund Halley noted that the stars Procyon, Sirius, and Arcturus had all changed their positions since the time the ancient Greek star catalogues were made. In the case of Procyon, this displacement was more than the diameter of the full moon. Because of the size and direction of this motion, Halley believed that it was not merely a relative motion, caused by some unknown movement of the earth, as had been the case with precession. Instead, he felt that these changes in the positions of Procyon, Sirius, and Arcturus were due to their actual movements through space. Halley expected that, in time, telescopic observations would reveal similar, though less dramatic, variations in the positions of many other stars. Such changes have been observed and are now known as proper motions. With the discovery of proper motions, the ancient concept of fixed stars was abandoned.

By the middle of the nineteenth century, observations of Procyon's proper-motion path showed that the star was not traveling in a straight line. Instead, a wavelike path was evident. According to Newton's laws of motion and gravitation, this form of proper motion suggested that a massive, faint companion star was orbiting Procyon, causing its path to vary over a period of about forty years. In 1896, the companion was discovered by James M. Schaeberle, using the new 36-inch refracting telescope at Lick Observatory, on Mount Hamilton, California. Subsequent measurements and calculations of the companion's orbit and luminosity showed that it is a white dwarf star.

Description·Procyon, at a distance of only 11.2 light-years, is one of the closest stars to the sun. Only fifteen stars are known to be nearer to us. However, less than half of these stars are visible to the naked eye. This implies that most of the bright stars seen in our night sky are considerably above average in intrinsic brightness.

Procyon is classified as a white subgiant or main-sequence star and is about 7 times brighter than the sun. Its surface temperature is 6,600°K. Ionized metals such as iron and calcium show the broadest absorption lines in the spectrum of Procyon.

Procyon's white dwarf companion is known as Procyon B; the two stars have a separation that varies between about 2 and 5 arcseconds. With a magnitude of only 11, the companion is lost in the glare of light from the brilliant primary star. Procyon B has a mass 65 percent that of the sun but a diameter only twice that of the earth. This combination results in an average density about 100,000 times that of the sun.

Canis Minor, *the Little Dog*

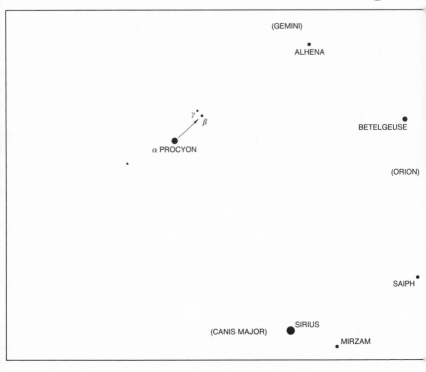

BRIGHT STAR OF CANIS MINOR

January 5	Procyon	0.4	white	subgiant – main sequence

KAY-niss MY-ner
C Mi
Canis Minoris

Evening Season•January – April

Sky Track•The Little Dog has an equatorial sky track, rising in the east, crossing the meridian halfway between the horizon and the zenith, and setting in the west.

Lore•The ancient Greeks described this constellation as Procyon, "before the dog." It was in Roman times that Procyon and its faint neighbors came to be known as the Little Dog or the Puppy. Canis Minor, along with Canis Major, represents one of Orion's hunting dogs. It has, at various times, been known as a dog belonging to

Actaeon, Diana, or Helen of Troy. In early Arabian sky lore the constellation marked one of the paws of a huge celestial lion.

Description•This small constellation with its single bright star is found just north of the celestial equator, on the eastern side of the Milky Way. The Little Dog is located to the southeast of the feet of Gemini, the Twins. The faint constellation Monoceros lies to the south, towards Canis Major.

Cancer, *the Crab*

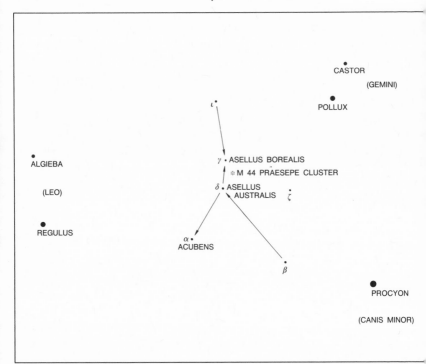

KAN-surr
Cnc
Cancri

Cancer is an inconspicuous zodiacal constellation that is located between Gemini and Leo. Its brightest star has a magnitude of only 3.5.

Evening Season•January – May

Sky Track•This constellation crosses the sky from the east point to the west and passes the meridian approximately halfway between the horizon and zenith.

Lore•This Crab is said to have been the creature that was sent by Hera to harass Hercules while he was struggling with the nine-headed Hydra in the marsh of Lerna. Hercules crushed the Crab after it had bitten his toes. As a reward for the little creature's brave effort, Hera placed it in the zodiac between Gemini and Leo.

Description•This is the faintest zodiacal constellation, yet it contains an object of considerable interest to naked-eye and binocular observers. The Praesepe (pray-SEE-pee) is one of the nearest star clusters to earth, with a distance estimated to be about 530 light-years.

The cluster is also known as the Beehive and it appears as a hazy spot to the unaided eye, with an apparent diameter about four times that of the full moon. Binoculars or low-power telescopes reveal the stellar nature of this group, which contains several hundred stars.

It is said that if the Praesepe cluster is invisible in an otherwise apparently clear sky, one may expect precipitation within 24 hours. Evidently, fine particles of ice form high in the atmosphere in front of an advancing storm system and obscure the Praesepe's delicate light before any clouds arrive to obscure the brighter stars.

The Praesepe star cluster is located about 40% of the angular distance from Pollux to the star Regulus.

Mirzam

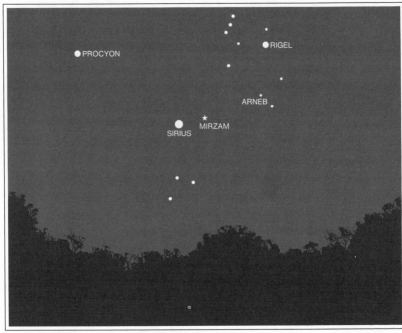

| Southeast | South-Southeast | South |

Mirzam	2.0 magnitude	−4.8 absolute
MIRR-zam	blue-white B1	magnitude
Beta, β	giant II–III	6h 23m −17° 57′
Canis Majoris,	700 light-years	February 11 transit
page 76	−0.013 −0.004 +34	STR 461

Location•Mirzam is a second-magnitude star located 6 degrees west of brilliant Sirius. It is 12 degrees east of the star Arneb, which is in the constellation Lepus.

Meaning•The name Mirzam is derived from its ancient description as the Preceding One. Mirzam appears to travel westward across the sky leading, or preceding, Sirius.

Lore•On traditional star maps, Mirzam generally marks one of the Large Dog's forefeet.

History•In 1908, it was found that the spectrum of Mirzam varies in a way indicating that the star expands and contracts slightly over a period of about six hours. Earlier, in 1902, Edwin B. Frost, director of the Yerkes Observatory, had discovered similar variations in the star Beta Cephei.

Description•Mirzam and Beta Cephei are members of a class of pulsating blue-white giant stars known as Beta Cephei variables, for the first star of this type to be discovered. These stars are characterized by light variations of only about 0.1 magnitude. The time required for the changes in brightness ranges from four to six hours, depending on the individual star. Mirzam, for example, has a light variation from magnitude 1.93 to magnitude 2.0, over an interval of about six hours. The designation of Mirzam as Beta Canis Majoris is favored by professional astronomers; because Mirzam is the brightest example of its class of variables, such variables are also known as Beta Canis Majoris stars.

Neutral helium and hydrogen are the most distinct absorption lines in the spectrum of Mirzam. The helium lines are wider than those of supergiant stars of spectral class B1, and are narrower than such lines seen in main-sequence stars of this class.

The proper motion of Mirzam is 0.013 arcsecond per year towards the west, and 0.004 arcsecond per year towards the south. The star moves 34 kilometers further from the earth every second.

The rate at which a star progresses in its evolution is mainly dependent on the amount of mass that it contains. Mirzam, with a mass about 10 times that of the sun, proceeds through its evolution hundreds of times faster than does the sun. The sun is believed to have a life expectancy of about 10 billion years; studies of meteorites believed to have been formed in the early days of the solar system indicate a present age for the sun of about 4.5 billion years. In contrast, massive Mirzam, with an age estimated to be in the order of magnitude of only about 15 million years, is believed to be pulsating due to the advent of the final stages in its life as a star. A star such as Mirzam is estimated to require an additional 4 million years to complete its evolution, possibly exploding as a supernova. These massive stars consume hydrogen fuel at a furious rate, just as a great bonfire may devour in a few hours an entire winter's supply of wood for a fireplace.

Canis Major, *the Large Dog*

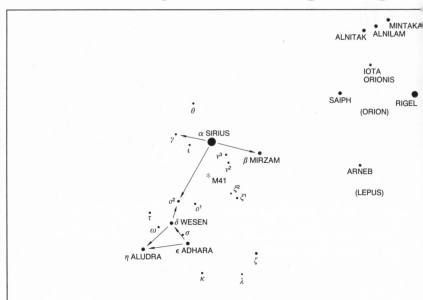

BRIGHT STARS OF CANIS MAJOR

January 16	Mirzam	2.0	blue-white	giant
January 19	Sirius	−1.5	blue-white	main sequence
February 20	Adhara	1.5	blue-white	giant
February 22	Wesen	1.8	white	supergiant
February 26	Aludra	2.4	blue-white	supergiant

KAY-niss MAY-jer
C Ma
Canis Majoris

Evening Season·January – March

Sky Track·This constellation rises in the southeast and sets in the southwest. Most of its bright stars are introduced in this guide as they transit the meridian, high point of their passage across the sky. Nevertheless, due to their southerly declination, they are seen closer to the horizon than most stars featured on the Sky Screen.

Lore•In ancient times, this constellation assumed the name Dog, title of its brightest star, Sirius. It has traditionally been described as the Dog of Orion. Other titles were the Hound of Actaeon, or the Dog of Procris, one of Diana's nymphs. In Scandinavia Sirius and its neighbors were associated with the dog Greip, in the Sigurd myth.

Description•Canis Major lies along the western side of the Milky Way, and is visible low in the southern sky.

The Dog's hips are marked by a small triangle formed by the second-magnitude stars Adhara, Aludra, and Wesen.

Dubhe

January 17

North	North-Northeast	Northeast

Dubhe	1.8/1.9 magnitude	−0.7 absolute
DUBB-ee	yellow K0	magnitude
Alpha, α	giant III	11h 04m +61° 45′
Ursae Majoris,	105 light-years	April 23 transit
page 80	−0.118 −0.071 −9	STR 465

Location•Dubhe marks the top front of the bowl of the Big Dipper.

Meaning•The name Dubhe is from the Arabian description of its location in the Back of the Greater Bear.

Lore•Hindu astronomers knew Dubhe as one of Seven Sages, which were represented by stars of the Big Dipper. Dubhe and its neighbor Merak, located 5 degrees to the south, are known as the Pointers. Follow a line from Merak through Dubhe and continue about 30 degrees further to Polaris.

History•Dubhe is a very close double star. A fifth-magnitude companion was discovered by Sherburne W. Burnham in 1889, using the 36-inch refractor at Lick Observatory. Burnham discovered 1,340 double stars and in 1906 published his *General Catalogue of Double Stars,* which described 13,665 of these objects.

Description • Dubhe is a yellow giant star located 105 light-years from the earth. It is about 110 times brighter than the sun. Dubhe is the only star of the seven in the Big Dipper that is not blue or blue-white. Alkaid, at the tip of the Dipper's handle, and Dubhe are the only Dipper stars that are not members of the Ursa Major moving cluster of stars. This cluster contains about twenty stars moving together through space at an average distance from the sun of 75 light-years and visible in the northern part of our sky.

The close fifth-magnitude companion discovered by Burnham is less than 1 arcsecond distant. In descriptions of double stars, the brighter of the two is called the primary or component A. The fainter star is known as the secondary, the companion, or component B. Dubhe and its companion are very difficult to resolve even with a large telescope.

Observers using a telescope may see a seventh-magnitude neighbor star about one-fifth the diameter of the full moon from Dubhe. This neighbor and the two components of Dubhe have similar proper motions.

Nearby Features

M81 (9h 54m +69° 09'), M82 (9h 54m +69° 47')

The interesting galaxies M81 and M82 are located about 10 degrees beyond Dubhe along a line extending from Phecda through Dubhe.

M81 has a magnitude of 8 and is classified as a spiral galaxy because of the beautiful and distinctive set of curving arms that appear to unwind from its nucleus. It is the brightest galaxy in a small group that lies about 10 million light-years from the earth.

M82 is another member of this group. Although its narrow oval shape suggests a spiral galaxy inclined to our line of sight, M82 shows no spiral arm structure and, unlike M81, cannot be resolved into stars. It is a source of radio signals and shows evidence of having a strong magnetic field. A set of filaments has been photographed extending outward from its core, apparently following lines of magnetic force.

See the following constellation map.

Ursa Major, *the Great Bear* (The Big Dipper)

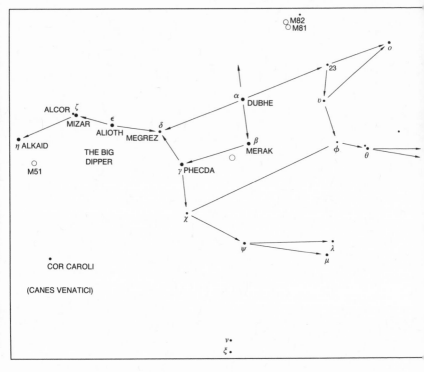

BRIGHT STARS OF URSA MAJOR

January 17	Dubhe	1.8	yellow	giant
January 22	Merak	2.4	blue-white	main sequence
February 5	Phecda	2.4	blue-white	main sequence
February 19	Alioth	1.8	blue-white	peculiar
February 28	Mizar	2.1	blue	main sequence
March 9	Alkaid	1.9	blue-white	main sequence

URR-sah MAY-jer
U Ma
Ursae Majoris

Evening Season·Ursa Major is easily seen during evenings in all months except October and November, when it is close to the horizon.

Sky Track•Circumpolar constellations such as Ursa Major, Ursa Minor, and Cassiopeia follow counterclockwise paths around the North Celestial Pole and do not set beneath the horizon.

Lore•Ancient Egyptians saw the seven stars of the Big Dipper as representing a bull's thigh or foreleg.

The Great Bear was one of the first constellations described in Greek literature, having been mentioned in *The Odyssey* of the ninth century B.C. This Bear was later related to the legend of Callisto and Arcas. Callisto was said to have been a beautiful princess of the kingdom of Arcadia with whom Zeus had fallen in love. Soon after Callisto's son Arcas was born, Hera, furious with jealousy, turned Callisto into a bear doomed to prowl the forests. One day years later, when Arcas had grown and was out hunting, Hera brought Callisto before him as prey. Zeus intervened, placing Callisto safely in the heavens as the Great Bear. Later, Arcas was put into the sky as the Little Bear. Hera, frustrated in her revenge, encouraged Poseidon to forbid the celestial Bears from dipping beneath the horizon, as do most other groups of stars. Thus the circumpolar nature of these constellations was explained. The extraordinarily long tails were formed, some say, when Zeus used them to swing the Bears around before slinging them into the sky. In this process, the tails were stretched far beyond normal lengths.

In parts of North America, stars of Ursa Major were also associated with a bear. Usually, the handle of the Big Dipper represented Three Hunters, or a Hunter and Two Dogs, pursuing the Bear, which was portrayed by the Dipper's bowl. The fourth-magnitude star Alcor, companion to Mizar in the handle, marked a pot to be used to cook the Bear. The Housatonic Indians of New England told that this Bear hunt took place each year between spring and autumn, at which time the Bear was finally wounded and its blood fell, giving color to the autumn leaves.

Chinese astronomers described the Dipper's bowl as the Northern Basket or Container, while in India the Dipper's stars were called the Seven Wise Men or Sages.

In England, the constellation's brightest stars have been known as King Arthur's Chariot, Charles's Wagon, the Butcher's Cleaver, and most generally, the Plough.

Description•The stars of the Big Dipper form a striking asterism, very pleasing to the eye and useful for showing directions in the sky.

Dubhe and Merak, the stars marking the front of the Dipper's bowl, help point us towards the North Star, located about 30 degrees from Dubhe. The constellation Leo is found 40 degrees south of the Dipper's bowl, and the graceful curve of the Dipper's handle leads towards the star Arcturus, a brilliant feature of springtime and summer skies.

Sirius

Southeast	South-Southeast	South

Sirius	−1.5 magnitude	1.4 absolute
SEAR-ee-us	blue-white A1	magnitude
Alpha, α	main sequence V	6h 45m −16° 43′
Canis Majoris,	8.6 light-years	February 16 transit
page 76	−0.545 −1.211 −8	STR 474

Location·Sirius is the brightest star in the night sky and is an unmistakable sight, located in the constellation Canis Major 25 degrees to the southeast of the Belt of Orion. It is nearly 10 times brighter than the average first-magnitude star.

Meaning·The name is from the Greek and means "the trembling one." Sirius "twinkles" so visibly because it is seen low in the sky where the atmosphere causes considerable disturbance to the star's brilliant light.

Lore·In ancient Egypt, Sirius was worshiped as the Nile Star. Its appearance in the predawn sky in late June, during the period around 3000 B.C., heralded the beginning of the annual Nile flood, which rejuvenated the soil of that great river valley. Sky-watchers in the ancient Middle East described Sirius as the Leader, referring to its dominant stature in the night sky.

The sun and Sirius are closest to each other in the sky during the hottest part of summer. In the Middle Ages it was believed that the combined light of these two objects produced that period of excessive heat. Since Sirius was also known as the Dog Star, the peak weeks of summer's heat came to be known as the Dog Days.

History•Between 1834 and 1844, Friedrich W. Bessel, using a 6¼-inch refracting telescope at the Königsberg Observatory, discovered that the proper motion of Sirius followed a wavelike path. On the evening of January 31, 1862, Alvan Graham Clark, Alvan Clark's younger son, discovered the faint companion to Sirius responsible for the wavy motion of the star. At the time, he was testing an 18½-inch telescope lens at his father's optical plant in Cambridge, Massachusetts. As a twelve-year-old, Alvan Graham Clark had kindled his father's interest in telescope making by melting down a broken bell from Phillips Academy, Andover, in order to cast a Newtonian mirror blank.

Description•Sirius has an apparent magnitude of -1.5 and a distance of 8.6 light-years. It is a blue-white main-sequence star with a surface temperature of about 10,200°K. Sirius has a luminosity 22 times that of the sun, but its apparent brilliance results from its relatively small distance from earth, rather than from a great intrinsic brightness.

Since Sirius is a member of a binary system, with a white dwarf companion, and is comparatively close to earth, it is possible to make a rather accurate estimate of the star's mass, which is calculated to be about 2.35 times that of the sun. The diameter of Sirius is estimated at about 1.8 times the diameter of the sun.

The companion discovered by Clark is known as Sirius B, and it was the first example of a white dwarf star ever found. It is sometimes called the Pup, and it has a magnitude of 8.5 and a surface temperature about the same as that of the primary. The white dwarf is very difficult to see because of its low brightness and the overpowering glare of Sirius A.

Maximum separation between Sirius A and B is about 11 arcseconds, which occurred in 1972. The year 1993 marks the closest approach, with a separation of just 2.5 arcseconds. The orbital period of the two components is about fifty years.

The average density of the white dwarf Sirius B is about 125,000 times that of water. A piece of this star's material the size of a tennis ball would weigh about as much as two full-grown elephants.

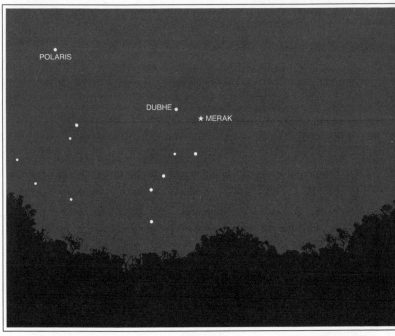

North-Northeast	Northeast	East-Northeast

Merak	2.4 magnitude	1.0 absolute
ME-rak	blue-white A1	magnitude
Beta, β	main sequence V	11h 02m +56° 23′
Ursae Majoris,	60 light-years	April 23 transit
page 80	+0.081 +0.029 −12	STR 483

Location•Merak is 5 degrees south of Dubhe. It is at the lower front of the Bowl of the Big Dipper.

Meaning•The name is from a description of this star as the Loins of the Bear.

Lore•Merak and Dubhe are the very helpful Pointer stars that aim our attention to the North Star. As with the other Dipper stars, Merak was one of the Seven Sages of India.

History•By 1869, proper-motion measurements showed that Merak and four other Dipper stars were moving in the same direction across the celestial sphere at the same speed. In 1872, the English astronomer William Huggins measured the Doppler shifts in spectral lines of the Big Dipper stars. From these observations, he was able to determine the radial velocities of these stars. Five Dipper stars have similar radial velocities as well as proper motions. Radial velocity indicates the speed at which a star is moving either towards or away from the sun. When combined, radial velocity and proper motion fully describe a star's true motion through space, which means that five of the Big Dipper stars are actually moving through space together as members of a star cluster. About twenty stars are known to be members of the Ursa Major cluster.

Description•Merak is a blue-white main-sequence star about 60 light-years from the earth. It is 33 times brighter than the sun and has a surface temperature of 10,200°K. Merak's spectrum displays very strong lines of hydrogen. These lines are much broader in main-sequence stars of class A1 than they are in giants and supergiants of this class. The proper motion is east 0.081, and north 0.029 arcsecond per year. The proper motion of Merak is 12 kilometers per second towards earth.

Nearby Feature

M97: THE OWL NEBULA (11h 14m +55° 08′)

In 1781, the astronomer Pierre Méchain, a colleague of Charles Messier's at the Paris Observatory, discovered a curious object about 2 degrees southwest of Merak. Messier included it in his list of nonstellar objects. It came to be known as M97 or the Owl Nebula because of two circular markings in the nebula that look like the eyes of an owl. Located at a distance of 12,000 light-years, the Owl is an example of a planetary nebula. Planetary nebulae resemble a planet when seen through a telescope but are actually shells of gas surrounding stars that are the later stages of their evolution. The Owl Nebula has a magnitude of only 11, and a telescope at least 8 inches in diameter is needed to observe it and to resolve the "eyes."

The Bright Stars of Winter

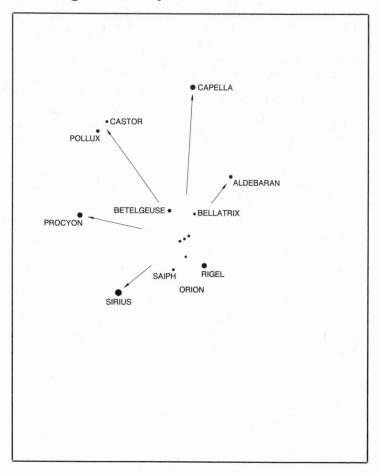

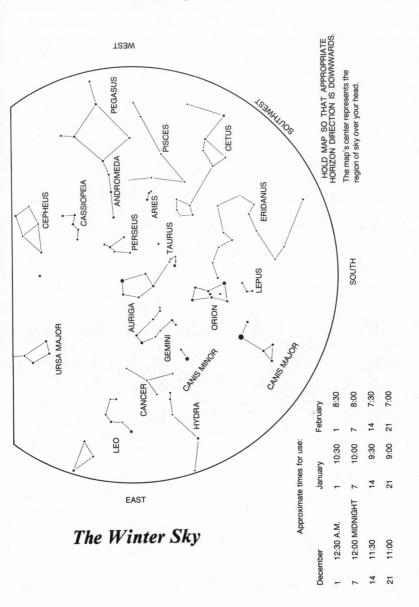

The Winter Sky

HOLD MAP SO THAT APPROPRIATE
HORIZON DIRECTION IS DOWNWARDS.

The map's center represents the
region of sky over your head.

Approximate times for use:

December		January		February	
1	12:30 A.M.	1	10:30	1	8:30
7	12:00 MIDNIGHT	7	10:00	7	8:00
14	11:30	14	9:30	14	7:30
21	11:00	21	9:00	21	7:00

Algieba   February 3

East-Northeast	East	East-Southeast
Algieba al-GEE-bah Gamma, γ Leonis, page 90	1.9/2.2 magnitude yellow K0 giant III 110 light-years +0.037 −0.151 −37	−0.4 absolute magnitude 10h 20m + 19° 51′ April 12 transit STR 532

Location • Algieba is a second-magnitude star located 8 degrees to the north of first-magnitude Regulus in the constellation Leo, the Lion. Algieba and Regulus are the brightest stars in the lovely asterism called the Sickle of Leo.

Meaning•The name is from an ancient description of this star's role in marking the Brow of Leo.

Lore•The association of Algieba with the Lion's Brow does not correspond to traditional Western portrayals of the constellation figure. In such representations, this star is shown in the Lion's mane.

History•In 1782, William Herschel discovered that Algieba was a double star. By 1831, Wilhelm von Struve at the Dorpat Observatory in Estonia had recorded a series of position measurements of the two component stars of Algieba. The companion's position changes over the years indicated that it was following an elliptical orbit around the primary star. As a result, the components of Algieba could be considered members of a gravity-bound binary-star system.

Description•The two components of the Algieba binary system present a fine sight in telescopes 6 inches in diameter and larger, using magnifications of 200 × or more. The stars are both yellow giants and have a separation of about 2.5 arcseconds. The magnitudes are 2.2 and 3.5 which blend to give a combined magnitude of 1.9. At a distance of 110 light-years, the angular separation of the pair corresponds to an actual distance of about four light-hours, nearly the same as the distance between the sun and the planet Neptune. Since the discovery by Herschel, only a small fraction of the orbital path has been observed. Therefore, accurate estimates of the period are difficult to obtain. Estimates of the time needed to complete one orbit range between 407 and 619 years.

The brighter component of the Algieba system is slightly variable in brightness, with a range of 0.15 magnitude.

Nearby Feature

40 LEONIS (10h 20m + 19° 28′)

Observers using binoculars can easily see the star 40 Leonis (Flamsteed number 40), located about 22 minutes of arc to the south of Algieba. Remembering that Algieba is a famous double star, you might at first think that you have resolved its components. However, 40 Leonis is over 500 times farther from Algieba, in angular measure, than is the orbital companion. The star 40 Leonis is a white giant at a distance of 65 light-years, and it is unrelated to Algieba.

The designation 40 Leonis indicates that this was the fortieth star of the constellation Leo to be catalogued by John Flamsteed in his *Historia Coelestis Britannica.* Many stars too faint to have received either a proper name or a Bayer Greek-letter designation are identified in this way.

Leo, *the Lion*

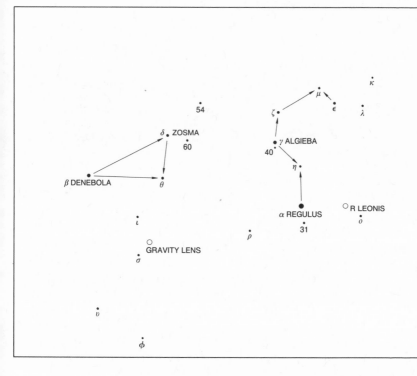

BRIGHT STARS OF LEO

February 3	Algieba	1.9	yellow	giant
February 6	Regulus	1.4	blue-white	main sequence
February 17	Zosma	2.6	blue-white	main sequence
March 1	Denebola	2.1	blue-white	main sequence

LEE-oh
Leo
Leonis

Evening Season•February – June

Sky Track•The stars of this zodiacal constellation rise in the east, pass the meridian high in the south, and set towards the west.

Lore•The Dendera zodiac or planisphere is a large sandstone medallion that shows many ancient constellations and asterisms. It was discovered on a ceiling in 1799 by one of Napoleon's officers in the temple of Isis at Dendera, which is located on the Nile about 60 kilometers north of Luxor. This star map is believed to date from the time of Cleopatra in the first century B.C. and to depict sky figures that were known in Egypt at that time, both indigenous as well as those borrowed from Mesopotamia and Greece.

The Dendera planisphere shows a lion in the part of the sky that we associate with Leo. Astrophysicist Owen Gingerich believes to be quite plausible the assertion, made by Egyptologist Virginia Lee Davis, that ancient Egyptians also saw a lion in these stars.

Ancient portrayals of a lion in this part of the sky may stem from the fact that, between about 2000 B.C. and the time of the Dendera planisphere, the sun occupied this portion of the zodiac during the hottest part of the summer. The fierceness of the midsummer sun may have been related to that of a lion. Furthermore, it has been said that at that time of the year, prides of lions migrated to the banks of the Nile to find relief from the oppressive heat. In Mesopotamia, the stars of Leo were shown as a shining disk or as a great fire, another metaphor for the sun's power during July and August.

In Greek and Roman sky lore, Leo was associated with the mighty Nemean Lion, the first challenge to Hercules as part of his Twelve Labors.

The constellation is said to have represented a sign of the tribe of Judah, the Lion having been assigned to Judah by Jacob as related in Genesis 49:9.

A horse was shown at this position of the Chinese zodiac until the sixteenth century, when the Lion was adopted.

Description•The first-magnitude star Regulus and the rest of the sickle that marks the head of the Lion are the most visually appealing parts of this constellation. A triangle of stars, located about 20 degrees to the east of Regulus, marks the hind portion of the Lion.

The Realm of the Galaxies extends from Leo into the constellation Virgo. This is the portion of sky that lies in the direction of the great Virgo Cluster of galaxies, the nearest large galaxy cluster to the Milky Way. The Virgo Cluster is centered about 50 million light-years from the earth, which is about 700,000 times farther from us than are the bright stars of Leo, neighbors of the sun in the Milky Way Galaxy.

Phecda February 5

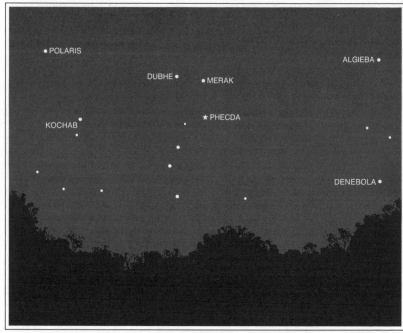

North-Northeast　　　　**Northeast**　　　　**East-Northeast**

Phecda	2.4 magnitude	0.5 absolute
FECK-dah	blue-white A0	magnitude
Gamma, γ	main sequence V	11h 54m +53° 42′
Ursa Majoris, page	80 light-years	May 6 transit
80	+0.093 +0.007 −13	STR 540

Location•Phecda is one of the four stars that outline the bowl of the Big Dipper. It is at the lower back of the bowl, diagonally across from Dubhe.

Meaning•The name is derived from a description of the Thigh, Phecda's position in the ancient figure of the Great Bear.

Lore•Phecda represented another of the Armillary Spheres of China and the Sages of ancient India, identified with the seven stars of our Big Dipper.

Description•Phecda is a blue-white main-sequence star located 80 light-years from the earth. It has 50 times the luminosity of the sun and a surface temperature of about 9,900°K.

The spectrum of Phecda contains very prominent hydrogen absorption lines. The width of these lines serves as a criterion of luminosity category for class A0 stars. Hydrogen lines in main-sequence stars such as Phecda are considerably broader than those seen in giants and supergiants of this class. Every year Phecda moves 0.093 arcsecond towards the east and 0.007 arcsecond northwards. Every second the distance between Phecda and the sun decreases by 13 kilometers.

Main-sequence stars of Phecda's type are estimated to have a mass and diameter about 2.5 times that of the sun. Such stars are believed to have main-sequence life expectancies in the order of magnitude of about 500 million years. (This means the actual age is probably closer to 500 million years than to either 50 million years or 5 billion years). After their time on the main sequence, stars expand and become giants, reaching the final stages of their evolution.

Regulus February 6

East-Northeast	East	East-Southeast

Regulus REG-you-luss Alpha, α Leonis, page 90	1.4 magnitude blue-white B7 main sequence V 85 light-years −0.249 +0.003 +6	−0.6 absolute magnitude 10h 08m +11° 58′ April 9 transit STR 543

Location•Regulus is 50 degrees south of the bowl of the Big Dipper and 40 degrees east of Procyon. It marks the base of the Sickle of Leo.

Meaning•The name is said to have originated with Copernicus and means the Little King. It is a diminutive form of the star's ancient title.

Lore•In the Sumerian civilization, Regulus was known as the Star of the King. In many parts of the world, this star, along with Aldebaran, Antares, and Fomalhaut, were known as Royal Stars. Such distinction was probably conferred because these stars lie near the ecliptic and are, in turn, prominent features of each of the four seasons.

Along the Euphrates, Regulus was known as the Flame or the Red

Fire. Through the ancient world it was believed that this star made a contribution to the heat of summer. About 2300 B.C., the summer solstice was located near Regulus. Therefore, around that period in history, the sun was located near Regulus at the start of summer, and the combined heat of the sun and the star was believed to produce the excessive heat of that season. Sirius, the brightest star in the night sky, later acquired a similar reputation when the precessional motion of earth's axis shifted the site of the summer solstice to the region of the ecliptic in the vicinity of this star.

A popular title for Regulus since classical times has been the Lion's Heart.

Ancient Chinese astronomers knew Regulus as the Bird Star.

History•Due to its brightness and proximity to the ecliptic, Regulus has long served as a celestial benchmark to which positions of the sun, moon, and planets have been compared.

Description•Regulus is a blue-white first-magnitude star. It lies at a distance of about 85 light-years and is 150 times more luminous than the sun. Regulus has a surface temperature of about 12,500°K. A star of this type is estimated to have a mass about three times that of the sun and a diameter three times greater, with a main-sequence life expectancy in the order of approximately 200 million years. Spectral class B stars such as Regulus show characteristically strong spectrum lines of neutral helium and for this reason they are sometimes known as helium stars.

There is an eighth-magnitude companion to Regulus at a distance of 177 arcseconds with an angular separation equal to about one-tenth the apparent diameter of the full moon. The companion to Regulus is itself a double star, having components of magnitudes 8 and 13, with a separation of 2.6 arcseconds.

Nearby Feature

VARIABLE STAR: R LEONIS (9h 48m +11° 56′)

R Leonis is one of the most frequently observed variable stars, and it is located about 5 degrees to the west of Regulus. It is a red giant, at a distance of about 600 light-years from earth. The brightness changes of R Leonis result from pulsations of this star over a period of 312 days. During that time the magnitude ranges from about 5 to 10. It is an example of a group of stars known as long-term variables.

Alphard

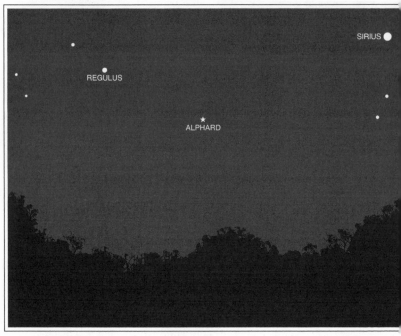

East-Southeast	Southeast	South-Southeast
Alphard AL-fard Alpha, α Hydrae, page 98	2.0 magnitude yellow K4 giant III 100 light-years $-0.018 + 0.028 - 4$	-0.4 absolute magnitude 9h 28m $-8°$ 40′ March 30 transit STR 583

Location•Alphard has a magnitude of 2.0 and may be found along a line extended from the tip of the handle of the Big Dipper onward past Regulus and 25 degrees farther towards the southwest. It is also on a line from Alhena through Procyon and 30 degrees beyond. There are no other bright stars in its immediate vicinity.

Meaning•The name means the Solitary One of the Serpent. Ulug-Beg called this star the Serpent's Neck.

Lore•In China, Alphard was the brightest star in a large asterism known as the Red Bird. Traditional maps usually show it marking the heart of a large serpent or sea snake.

Description•Alphard is a yellow second-magnitude star. It is a giant located about 100 light-years from earth. It has a surface temperature of nearly 4,000°K and is about 120 times brighter than the sun.

There is a tenth-magnitude optical companion at a distance of 283 arcseconds. In stars such as Alphard the most prominent spectral lines are those of neutral metals. These lines are most intense in this class of stars.

Alphard has a proper motion of 0.018 arcsecond to the west and 0.028 arcsecond north each year. The star's distance from the sun is decreasing at the rate of about four kilometers per second.

Nearby Feature

COSMIC BACKGROUND RADIATION

Cosmic background radiation is the stretched-out remnant of light from the primeval fireball that marked the origin of our universe. This radiation has proven to be extremely uniform in intensity and reaches us from all directions of the sky. In 1977 however, a "hot spot" in the CBR was discovered in the general direction of the constellations Hydra and Leo, to the east of Alphard. This phenomenon is attributed to the motion of the Milky Way Galaxy as the Galaxy orbits the center of the Virgo Supercluster of Galaxies. As a result of this galactic motion, the CBR in that direction is subject to a Doppler shift towards shorter (hotter) wavelengths.

Hydra, *the Water Snake*

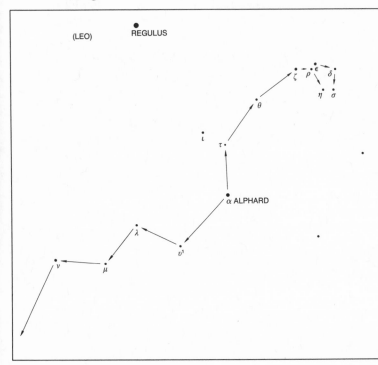

BRIGHT STAR OF HYDRA

February 16	Alphard	2.0	yellow	giant

HIGH-druh
Hya
Hydrae

Evening Season•April – June

Sky Track•Hydra is the longest constellation, stretching across nearly one-third of the celestial sphere. This constellation lies mainly to the south of the celestial equator, and it moves across the sky from the southeast to the southwest.

Lore•The ancient constellation of Hydra was associated in Greek mythology with the water snake carried by Corvus, the Crow, to Apollo as a scapegoat for the Crow's indolence.

Another Greek myth relates Hydra to the Serpent of Aetes, which guarded the Golden Fleece and was lulled to sleep by the sweet song of Medea.

The figure of the ancient Hydra was extended towards the east in about the tenth century A.D. and now reaches the constellation Libra. The prototype for this enormous figure was the legendary Kraken, an immense water snake said to inhabit the seas around Scandinavia.

Description•With the exception of Alphard, the stars of Hydra are of the third magnitude or fainter, and a clear night is needed to trace the entire figure.

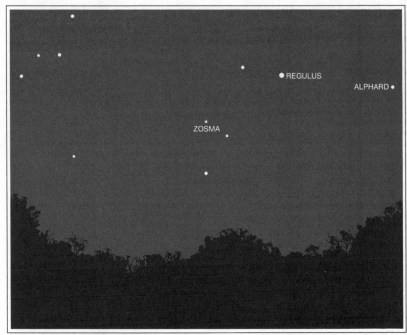

East-Northeast	East	East-Southeast

Zosma ZOSE-mah Delta, δ Leonis, page 90	2.6 magnitude blue-white A4 main sequence V 80 light-years +0.143 −0.135 −20	0.7 absolute magnitude 11h 14m +20° 31′ April 26 transit STR 584

Location•Zosma is 14 degrees east of Algieba. It is at the northern point of a triangle of stars that marks the hind portion of Leo.

Meaning•The name originally indicated the Star on the Back of Leo the Lion.

Lore•Zosma and Theta, 5 degrees to the south, are said to have been known to ancient Euphratean astronomers as the Oracle. More recently, they have been called the Two Little Ribs. Chinese astronomers called Zosma the Higher Minister of State.

History•On December 13, 1690, John Flamsteed was looking at stars near Zosma and recorded the position of an object that was much later determined to have been the planet Uranus. This is termed a prediscovery observation because Flamsteed did not comment on the object's unusual appearance or attempt to determine its true nature. Credit for this planet's discovery goes to William Herschel, who, in 1781, saw Uranus and called attention to its unstarlike appearance.

Description•Zosma is a blue-white main-sequence star with a magnitude of 2.6 and a distance of about 80 light-years from earth. Its surface temperature is approximately 8,600°K, and Zosma has 44 times the luminosity of the sun.

The spectrum of this star displays strong absorption lines produced by the element hydrogen. Other characteristic lines include those of various ionized metals, such as magnesium, iron, and titanium.

Stars such as Zosma are estimated to have masses about 2.3 times greater than the sun's, with diameters about twice as large. This type of star has a main-sequence life expectancy in the order of magnitude of about 500 million years, after which it evolves and becomes a giant.

Alioth

North-Northeast   Northeast   East-Northeast

Alioth	1.8 magnitude	−0.2 absolute
AL-ee-oth	blue-white A0	magnitude
Epsilon, ϵ	peculiar	12h 54m +55° 58′
Ursae Majoris,	80 light-years	May 21 transit
page 80	+0.109 −0.010 −9	STR 595

Location•Alioth is a second-magnitude star in the handle of the Big Dipper. It is the brightest star in the Dipper and is third in order from the end of the handle.

Meaning•The origin of the name Alioth may have come from an Arabian description of this star as the Gulf, possibly referring to its position near the middle of the Big Dipper.

Lore•According to American Indian legends, Alioth marked the first of three hunters stalking a bear, represented by the stars of the Dipper's bowl.

Alioth was identified with an ancient Chinese astronomical instrument called a traverse, which was used to measure star positions.

History•John Herschel became a celebrated astronomer like his father, William, and carried on studies of double stars, clusters, and nebulae. John extended his observations to include objects seen only from the southern hemisphere. In 1833, he traveled to the Cape of Good Hope and used his father's 20-foot telescope to make the first modern survey of celestial objects visible in the southern hemisphere.

In 1847, the same year that the results of his Cape observations were published, Herschel noted that Alioth had apparently faded slightly and appeared to be less bright than Alkaid. Alioth evidently soon returned to its more characteristic brightness range.

Description•With an average magnitude of about 1.8, Alioth is the brightest star in the constellation of Ursa Major, the Great Bear.

Alioth is an example of a type of pulsating variable star known as magnetic spectrum variables. Cor Caroli, in the constellation Canes Venatici, is the prototype of such stars.

With a surface temperature of about 9,900°K, Alioth has a blue-white color. Its magnitude varies between 1.76 and 1.79 over a period of 5.09 days. The star is located about 80 light-years from earth.

The spectrum of Alioth is classified as peculiar, in part due to the presence of unusually strong spectral lines from the elements europium and chromium.

Adhara

South-Southeast	South	South-Southwest

Adhara
a-DAY-rah
Epsilon, ϵ
Canis Majoris,
    page 76
	1.5 magnitude
	blue-white B2
	giant II
	650 light-years
	+0.001 +0.002 +27
		−5.0 absolute
		    magnitude
		6h 59m −28° 58′
		February 20 transit
		STR 599

Location•Adhara is found 13 degrees to the south of Sirius. It is at the southwestern corner of a small triangle of second-magnitude stars in Canis Major.

Meaning•The name is taken from a description in Arabian sky lore of Adhara and its neighbors, Aludra and Wesen, as the Maidens.

Lore•Adhara marks part of the hind portion of Canis Major, the Large Dog.

History•In 1850, an eighth-magnitude companion star to Adhara was discovered at the Cape Observatory in South Africa. The separation of 7.5 arcseconds has not changed since that time, and therefore these stars seem to be unrelated. The glare from Adhara makes it very difficult to see this companion.

Description•Adhara is a blue-white giant star at a distance of about 650 light-years from earth. It is 8,500 times brighter than the sun and has a surface temperature of about 23,000°K.

With an apparent magnitude of 1.5, Adhara is at the dividing line between first- and second-magnitude stars. First-magnitude stars range from 0.50 to 1.49; second-magnitude stars range from 1.50 to 2.49. Further categories by magnitude group stars in a similar way. This system of dividing ranges of magnitude makes Adhara one of the brightest-possible second-magnitude stars.

Dark absorption lines produced by the presence of neutral helium in this star's outer atmosphere characterize the spectrum of Adhara. Lines of ionized silicon, oxygen, and magnesium are also strong. The hydrogen lines of Adhara and other giants of class B2 are weaker than those seen in B2 stars of lower luminosity, i.e., subgiants and main-sequence stars. Supergiants of class B2, such as the star Chi^2 Orionis, have hydrogen lines even narrower than those seen in spectrum of Adhara.

The proper motion of Adhara, in arcsecond per year, is 0.001 eastward and 0.002 towards the north. Adhara's spectrum reveals a radial velocity of 27 kilometers per second away from the sun.

As the sun progresses along its 200 million year orbit around the center of the Milky Way Galaxy, it moves away from a point on the celestial sphere known as the "solar antapex." This point is located about 10 degrees of arc to the west of Adhara. As a result of this motion of the solar system, the distance between the sun and stars seen in this part of the sky is generally increasing. The radial velocity of Adhara, +27 kilometers per second, is a good example of this effect. Radial velocities of stars in the opposite part of the sky, around the constellations Lyra and Hercules, usually reveal decreasing distances from the sun, because the solar system is moving towards that direction.

Wesen   February 22

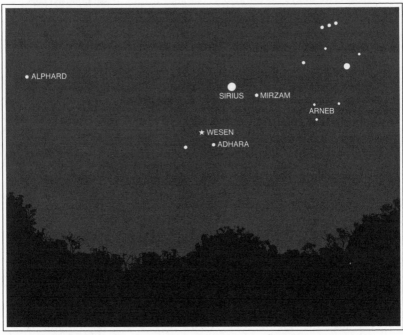

South-Southeast	South	South-Southwest

Wesen	1.8 magnitude	−7.3 absolute
WE-zen	white F8	magnitude
Delta, δ	supergiant Ia	7h 08m − 26° 24′
Canis Majoris,	2,000 light-years	February 22 transit
page 76	−0.008 +0.003 +34	STR 608

Location•Wesen is at the northern corner of the small triangle of second-magnitude stars in the lower part of Canis Major. It is about 3 degrees northeast of Adhara.

Meaning•The name Wesen was apparently applied to this star in the early part of the nineteenth century by the astronomer Giuseppi Piazzi. It represents the arbitrary adaptation and application of an ancient star name from the neighboring constellation of Columba to Delta Canis Majoris, which had no traditional proper name.

Lore•Wesen marks the location of the hips of the Large Dog, as shown in some portrayals of the constellation.

Description• At a distance of nearly 2,000 light-years, Wesen is one of the most remote stars visible to the naked eye. Estimates of the distances to stars so far from earth are possible with the help of spectral analysis. The apparent magnitudes of these stars are compared to absolute magnitudes characteristic of their special classes.

Wesen is a white supergiant star. It is one of the most luminous that can be seen with the naked eye and is about 70,000 times brighter than the sun. Its surface temperature is about 6,000°K. Ionized metals in the outer atmosphere of Wesen produce the most distinctive lines in the star's spectrum. Absorption lines of hydrogen and neutral metals are also evident. The proper motion of this star is small, 0.008 arcsecond to the west and 0.003 arcsecond towards the north, per year. Small values of proper motion are characteristic of stars such as Wesen that are thousands of light-years from the sun. The distance, proper motion, and radial velocity values of Wesen and those of the neighboring star Aludra are quite similar.

Aludra   February 26

South-Southeast	South	South-Southwest

Aludra
ah-LUD-rah
Eta, η
Canis Majoris,
    page 76
	2.4 magnitude
	blue-white B5
	supergiant Ia
	2,500 light-years
	−0.008 +0.002 +41
		−7.0 absolute
		    magnitude
		7h 24m −29° 18′
		February 26 transit
		STR 622

Location•Aludra is at the eastern corner of the triangle of second-magnitude stars found to the south of Sirius in Canis Major.

Meaning•The origin of this star name is uncertain.

Lore•Aludra is usually shown near to the tail end of the figure of the Large Dog.

Description•Blue-white supergiant stars such as Aludra are among the most luminous that we can see in the sky. Aludra is about 55,000 times brighter than the sun. Because of its intrinsic brilliance, Aludra is a feature of our winter evenings with a magnitude of 2.4, even at its distance of 2,500 light-years. If the sun were at this distance, it would have an apparent magnitude of about 14 and you would need a 12-inch telescope in order to see it.

Hydrogen and helium lines are prominent in Aludra's spectrum. The intensities and widths of various hydrogen lines provide the identification of Aludra as a type Ia supergiant. These lines are more intense and wider in giant and main-sequence stars of spectral class B5.

Aludra's proper motion, radial velocity, and distance reveal a commonality with Wesen. Both stars are members of a small cluster known as Collinder 121, whose members are related in terms of time and location of their formation.

Although Wesen and Aludra are near the star Adhara in the constellation Canis Major, Adhara's motions and distance from earth suggest that it is an outlying member of the Pleiades group of stars, with no actual association to either Wesen or Aludra.

Aludra has a seventh-magnitude optical, therefore unrelated, companion at a distance of 169 arcseconds.

Kochab  February 27

North   North-Northeast   Northeast

Kochab	2.1 magnitude	−0.5 absolute
KOE-cab	yellow K4	magnitude
Beta, β	giant III	14h 51m +74° 09′
Ursae Minoris,	110 light-years	June 20 transit
page 62	−0.035 +0.010 +17	STR 627

Location·Kochab is the brighter of two stars at the front of the Little Dipper's bowl. This second-magnitude star is 16 degrees from Polaris.

Meaning·The origin of the name is obscure. It has sometimes been associated with a Hebrew word for "star."

Lore·Kochab and its neighbor Pherkab served as twin pole stars during the period of classical antiquity. Their ancient title was the Guardians of the Pole.

History·Kochab and its neighbor were regarded as the pole stars from about 1500 B.C. until A.D. 500, when the role was gradually assumed by our Polaris, Alpha Ursae Majoris. This succession of pole stars is a result of earth's precessional motion.

Description• Kochab is the second-brightest star in the Little Dipper and has a magnitude of 2.1. It is a yellow giant at a distance of 110 light-years from earth. This star is 130 times more luminous than the sun, with a surface temperature of about 4,000°K.

Mizar February 28

North-Northeast	Northeast	East-Northeast

Mizar	2.1/3.0 magnitude	0.8 absolute
MY-zar	blue-white A2	magnitude
Zeta, ζ	main sequence V	13h 24m +54° 55′
Ursae Majoris,	90 light-years	May 29 transit
page 80	0.119 −0.025 −6	STR 628

Location•Mizar is the second-magnitude star that marks the bend of the Big Dipper's handle.

Meaning•The name Mizar was given to this star at the beginning of the nineteenth century. It was adapted from an Arabic word meaning "the apron."

Lore•In an American Indian legend, Mizar and its fourth-magnitude companion, Alcor, represented a hunter with a kettle, stalking a huge bear through the forest.

Another description for the Alcor-Mizar pair, and which has long been popular in England, is as the Horse and Rider.

In China, Alcor was known as the Supporting Star.

The pair has been long used as a test of eyesight, and the combination of star brightnesses and separation provides a good challenge to visual acumen.

History•Mizar is one of the finest examples of a multiple-star system. It consists of a 2.1-magnitude primary called Mizar A and a sixth-magnitude companion known as Mizar B, located about 14 arcseconds from the primary. This companion was discovered by Giovanni Battista Riccioli in 1650. James Bradley began to make systematic measurements of the components in 1755.

Mizar A, the primary component of the star system, was itself the first spectroscopic double to be discovered. In 1889, Edward C. Pickering, director of the Harvard College Observatory, found that the spectral lines of Mizar A are usually seen as pairs of lines. Furthermore, he found that the separation of these line pairs varied over a period of time, and at certain points the lines merged into single sets of lines. Pickering deduced that this effect was caused by Mizar A actually having two components so close together that telescopes could not visually resolve them into separate stars.

Description•Mizar A is composed of two nearly identical blue-white main-sequence stars each of magnitude 3.0 and having a separation of about 0.01 arcsecond. The period of this pair is about 21 days.

Mizar B, at a distance of 14 arcseconds from Mizar A, is believed to show evidence that it is composed of the blended light of three type-A main-sequence stars. Irregularities in the orbital path of the Mizar B system suggest the presence of yet another component, at a distance of 0.13 arcsecond from Mizar B, with an orbital period of 57 years.

Alcor has the same proper motion as the stars in the Mizar AB system and is believed to be a gravity-bound member of the system.

Mizar is an interesting star to observe in turn with the naked eye, binoculars, and a telescope. Try to find Alcor with your unaided eye and then look at this lovely pair with binoculars. Next, use a telescope to study the attractive components Mizar A and B.

Denebola

March 1

East-Northeast	East	East-Southeast

Denebola de-NEB-oh-la Beta, β Leonis, page 90	2.1 magnitude blue-white A3 main sequence V 42 light-years −0.497 −0.119 −0	1.6 absolute magnitude 11h 49m +14° 34′ May 5 transit STR 635

Location•Denebola is the easternmost and, at magnitude 2.1, the brightest of the stars forming the triangle marking the rump of Leo, the Lion. Denebola is 25 degrees to the east of Regulus.

Meaning•The name Denebola originated with a description of this star's position, at the Lion's tail.

Lore•Ulug-Beg commented that the appearance of Denebola was a sign of a change in the weather. The sharp cold of winter begins to leave us during the weeks when this star appears in the eastern evening sky. Months later, when Denebola sets in the west, the nights are getting longer and the hottest part of the year draws to a close.

In China, Denebola and several neighboring stars were known as the Seat of the Five Emperors.

On the subcontinent of India, this was called the Star of the Creating Mother Goddess.

History•Sherburne W. Burnham initiated the modern era in double-star study when in 1870 he began systematic observations using a 6-inch Clark refractor from the backyard of his Chicago home. He soon showed that there were a great many double stars yet to be discovered. Although he remained an amateur for most of his career, Burnham earned the use of many of America's largest telescopes, including those at the Lick and Yerkes observatories. In 1878, he discovered a thirteenth-magnitude companion to Denebola.

Description•Denebola is a blue-white main-sequence star at a distance of 42 light-years from earth. Its surface temperature is about 8,900°K, and the luminosity is 20 times that of the sun. Denebola is an example of spectral class A3, in which the spectral lines of hydrogen are strong. In addition, lines of singly ionized silicon, iron, and other elements are seen.

Main-sequence stars similar to Denebola are estimated to have masses and diameters about twice those of the sun. Such stars have a main-sequence life expectancy of approximately 3 billion years.

In addition to the optical companion discovered by Burnham at a distance of 77 arcseconds, Denebola has a sixteenth-magnitude binary companion 40 arcseconds distant.

Nearby Feature

M64: THE BLACKEYE GALAXY (12h 56m +21° 48′)

M64, known as the Blackeye Galaxy, is located about 18 degrees to the west of Denebola, in the constellation Coma Berenices. The name was inspired by a prominent dust lane in the middle of the galaxy. A 6-inch telescope and a clear, dark night are needed in order to see M64's dust spot. This galaxy is located about 22 million light-years from us.

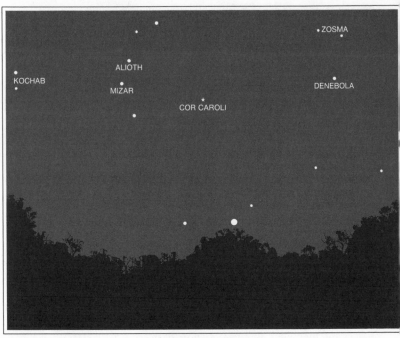

Northeast   **East-Northeast**   **East**

Cor Caroli	2.8/2.9 magnitude	0.1 absolute
core-CAROL-eye	blue-white A0	magnitude
Alpha, α	peculiar	12h 56m +38° 19′
Canum Venati-	120 light-years	May 22 transit
corum, page 118	−0.236 +0.052 −3	STR 643

Location•Cor Caroli is nestled under the curved handle of the Big Dipper, 17 degrees to the south of Alioth.

Meaning•One tale claims that this star received its name in 1725 from Edmund Halley. He is said to have named it Cor Caroli, "Heart of Charles," in honor of King Charles II of England. Deborah Jean Warner of the Smithsonian Institution attributes the name to a commemoration of Charles I.

Lore•Cor Caroli marks the collar of Chara, one of the dogs of the constellation Canes Venatici, the Hunting Dogs. The second dog, Asterion, is traditionally represented by a line of faint, fifth-magnitude stars 5 to 10 degrees east of Cor Caroli.

Description•Although it is one of the fainter stars described in *The Star Guide,* Cor Caroli is a favorite double star for observers with small telescopes. The components have magnitudes of 2.9 and 5.6, with a separation of about 19 arcseconds. The color of the primary is blue-white and the secondary appears yellow. The pair shows a lovely separation at 100×. The distance to these components is about 120 light-years.

The brighter star in the pair is known as Alpha^2 Canum Venaticorum. It is the best-known example of a type of star called a magnetic-spectrum variable. These are spectral-type-A stars that do not pulsate, unlike most other types of variable stars. They have very strong spectral lines of the elements strontium, silicon, chromium, or certain other elements. These lines change in strength over periods between one and twenty-five days. There is also a slight light variation, but this is less than a range of 0.1 magnitude. Alpha^2 has a period of variation of 5.5 days.

Such stars also have powerful magnetic fields, which vary over the same period of time as the spectral lines and visible light. Its magnetic field has a maximum intensity about 5,000 times that of the earth's, varying between northern and southern polarity. Some astronomers believe that, like earth, the magnetic axis is inclined somewhat with respect to the rotational axis. As the star rotates, first a north pole then a south pole faces earth. Perhaps different chemical elements gather at each pole, causing the variation in spectral lines as the star rotates.

Nearby Feature

Y CANUM VENATICORUM (12h 43m +45° 43′)

The star known as Y Canum Venaticorum is 7 degrees north of Cor Caroli. It has an interesting orange color and is a semiregular variable with a range from 5 to about 6.5 over 160 days. It is one of the coolest stars known, with a surface temperature of 2,600°K. The unusual color prompted Pietro Angelo Secchi to call it "La Superba." The star's relatively cool surface temperature allows carbon to exist in a molecular form; such stars are called carbon stars, due to their characteristic spectral lines of molecular carbon.

Arcturus, Spica, and Denebola form a sky pattern or "asterism" called the Spring Triangle. Addition of the star Cor Caroli forms a larger asterism known as the Diamond of Virgo.

See the following constellation map.

Canes Venatici, *the Hunting Dogs*

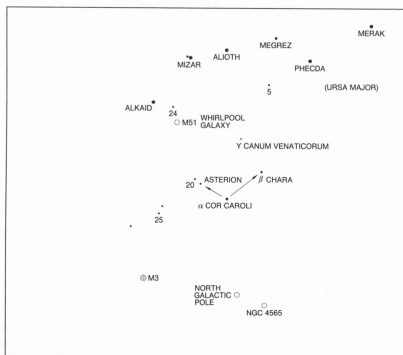

MERAK

MEGREZ

ALIOTH

MIZAR

PHECDA

5

(URSA MAJOR)

ALKAID

24

M51 WHIRLPOOL
GALAXY

Y CANUM VENATICORUM

20 ASTERION β CHARA

α COR CAROLI

25

M3

NORTH
GALACTIC
POLE

NGC 4565

BRIGHT STAR OF CANES VENATICI

March 3	Cor Caroli	2.8	blue-white	peculiar

KAY-nees vee-NAT-ih-sigh
CVn
Canum Venatacorum

Evening Season•March – August

Sky Track•This constellation passes through the sky from the northeast, moves across the meridian overhead, and sets in the northwest.

Lore•The group of stars that includes Cor Caroli was not known in Greece and Rome as a separate constellation. Ptolemy described them as a part of Ursa Major.

The Polish astronomer Johannes Hevelius of Danzig devised the constellation Canes Venatici, and it was first shown in his star atlas, called the *Prodromus,* published posthumously in 1690.

Cor Caroli and the fourth-magnitude star Chara mark the location of the Southern Hound, which is also known as Chara. The Northern Hound of Canes Venatici is called Asterion.

John Flamsteed included Canes Venatici and several other constellations invented by Hevelius in his star atlas, published in 1729. These constellations of Hevelius, unlike most that have been devised in modern times, were accepted into the astronomical literature.

Description • With the exception of Cor Caroli, the stars of Canes Venatici are all of the fourth magnitude or fainter. However, the constellation is located near the North Galactic Pole, found about 10 degrees south of Cor Caroli, and near one of the two most "transparent" parts of the Milky Way Galaxy, about 90 degrees from its congested plane. For this reason, the region of sky around Canes Venatici is relatively free from Milky Way dust. This provides a clear "porthole" through which we can glimpse galaxies beyond the Milky Way, including M51, the famous Whirlpool Galaxy.

Alkaid

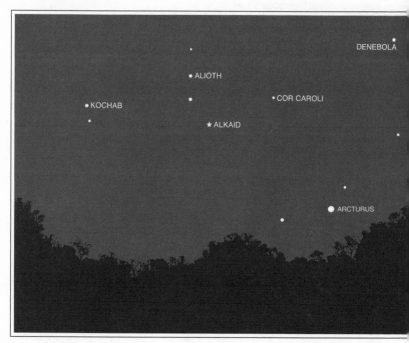

North-Northeast	**Northeast**	**East-Northeast**

Alkaid
al-KADE
Eta, η
Ursae Majoris,
   page 80
	1.9 magnitude
	blue-white B3
	main sequence V
	150 light-years
	$-0.126 -0.014 -11$
		-1.6 absolute
		   magnitude
		13h 48m $+49°$ 19′
		June 4 transit
		STR 668

Location•Alkaid is the star at the end of the Big Dipper's handle.

Meaning•The name indicated the Leader of the Procession, referring to the train of stars that forms the Big Dipper.

Lore•In ancient China, Alkaid was known as the Revolving Light. It represented one of the Seven Sages of India, which were each marked by a star in the Big Dipper.

History•Alkaid was one of the stars that was carefully observed by James Bradley during the eighteenth century, in his unsuccessful attempt to measure their distances. He tried to detect parallactic shifts in star positions as the earth moved from one side of the sun to the other during a six-month period. Bradley's instruments were, however, insufficient to show these shifts. The first measurement of stellar distance was not made until the next century, when Bessel determined the parallax of the star known as 61 Cygni.

Description•With a magnitude of 1.9, Alkaid is a blue-white main-sequence star located about 150 light-years from earth. It is seen to trail the other stars of the Big Dipper as they circle the celestial pole in a counterclockwise direction. Alkaid has a surface temperature of about 20,000°K and a luminosity approximately 350 times greater than the sun's.

Stars such as Alkaid are estimated to have masses about 3.5 times that of the sun and main-sequence life expectancies in the order of 100 million years.

Alkaid, along with Dubhe, are the only two Big Dipper stars that are not members of the Ursa Major star cluster.

Nearby Feature

M51: THE WHIRLPOOL GALAXY (13h 29m +47° 18′)

The spectacular Whirlpool Galaxy, M51, is located 3.5 degrees to the southwest of Alkaid. This galaxy has a magnitude of 8, and a small telescope will show it as a hazy spot when the night is clear and dark. Long photographic exposures made with large telescopes reveal M51's magnificent appearance.

Until the early part of the twentieth century, M51 and similar "nebulae" were considered to be either clouds of gas or unresolved groups of stars located within the Milky Way system of stars. During the 1920s, spiral "nebulae" such as M51 were conclusively shown to be separate and enormously distant systems of stars, dust, and gas, comparable to the entire Milky Way. The Whirlpool Galaxy, for example, is about 35 million light-years from earth, a distance about 350 times greater than the diameter of the Milky Way.

In 1845, William Parsons, the third earl of Rosse, discovered a delicate network of spiral filaments winding through M51. Parsons made these observations using a mammoth 72-inch reflecting telescope, which he had constructed on the grounds of his ancestral estate, Birr Castle, in Ireland. This instrument was known as the Leviathan of Parsonstown, and it was, by far, the world's largest telescope at that time. The discovery of spiral structure in M51 was one of the several major contributions to astronomy made with the 72-inch reflector. Its mounting and sighting devices were inadequate, and weather conditions also prevented full realization of this telescope's potential.

See the constellation maps of Ursa Major (page 80) and Canes Venatici (page 118).

Vindemiatrix

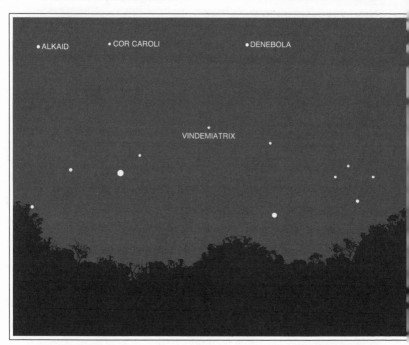

East-Northeast	East	East-Southeast

Vindemiatrix vin-DEE-my-A-trix Epsilon, ε Virginis, page 124	2.8 magnitude yellow-white G8 giant III 75 light-years −0.275 +0.017 −14	1.0 absolute magnitude 13h 2m + 10° 58′ May 8 transit STR 720

Location•Vindemiatrix is a third-magnitude star located about 20 degrees east of Denebola and 25 degrees to the south of Cor Caroli.

Meaning•The name is of Latin origin and means "the vine grower."

Lore•Vindemiatrix is an example of a star that served in ancient times as an indicator of the season. Its arrival in the predawn eastern sky heralded the time of the grape harvest.

Description•This is a yellow-white giant star, at a distance of 75 light-years from the earth. It has a luminosity 35 times greater than the sun's.

Nearby Feature

THE VIRGO CLUSTER

The great Virgo Cluster of galaxies covers a large portion of the sky between the constellations Leo and Virgo. This vast assembly of thousands of galaxies is centered about halfway between Vindemiatrix and Denebola.

The brightest galaxies in this cluster have apparent magnitudes of 8 or 9 and may be seen with small telescopes and binoculars as spots of haze if the sky is very clear and dark.

The giant elliptical galaxy, M87 (12h 30m + 12° 30′), is one of the brightest galaxies in the Virgo Cluster and is located 8 degrees west of Vindemiatrix. It has a magnitude of 9.2 and a distance from earth of about 45 million light-years. M87 typifies the giant elliptical galaxies found near the centers of rich clusters of galaxies. See the following constellation map.

The Virgo Cluster is near the middle of an even larger collection of galaxies known as the Local Supercluster. The Milky Way and Andromeda galaxies are members of the small Local Group of about twenty galaxies that also belongs to this supercluster.

This is the time of year when the sun, moving along the ecliptic towards the east, reaches the celestial equator and the zero-hour circle of right ascension at the point known as the vernal equinox. With this event, spring begins in the northern hemisphere and autumn in the southern hemisphere. At the start of our spring, the sun is at a direction on the celestial sphere about 15 degrees to the south of the star Algenib in the constellation Pegasus. Each year, the precession of the earth's axis carries the location of the vernal equinox towards the west, along the ecliptic, relative to the stars, an angular distance of about 50 arcseconds.

Virgo, *the Virgin*

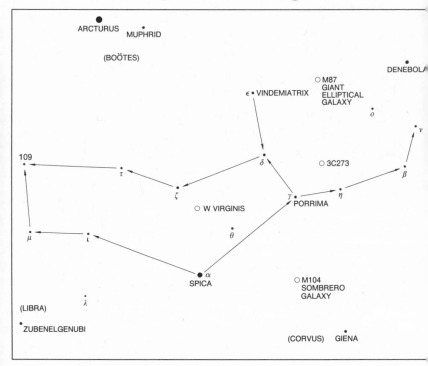

BRIGHT STARS OF VIRGO

March 23	Vindemiatrix	2.8	yellow-white	giant
March 28	Porrima	2.8	blue-white	main sequence
April 21	Spica	1.0	blue-white	main sequence

VER-go
Vir
Virginis

Evening Season•April–July

Sky Track•The stars of Virgo lie along the celestial equator. The constellation rises directly to the east, crosses the meridian halfway between the horizon and the zenith, and sets near the western point of the horizon.

Lore•The legends of Virgo are closely tied to the images of spring-time. Suggestions have been made that these associations arose at the time when the vernal equinox was located amidst the stars of Virgo, about 14,000 years ago. Such an origin would attest to the remarkable durability of an oral tradition, but this speculation has never been verified.

Mesopotamian sky-watchers associated Virgo's stars with Ishtar, their goddess of love and fertility. Under the name Ashtoreth, this goddess was mentioned in the Bible, I Kings 11:5 and 33. Ishtar was the prototype for Aphrodite in Greek mythology and Venus in Roman, and she may even have been the source for our word *star*. Phoenicians used the name form Astarte, which the seventh-century English scholar the Venerable Bede associated with Eastre, the Saxon goddess of spring. It was Eastre who provided the English name for the Christian festival of the Resurrection.

In ancient Greece, the stars of Virgo were known as Persephone, the maiden of spring and daughter of Demeter, the goddess of grain and the harvest. Virgo was also seen as a representation of Demeter, carrying a sheaf of grain under one arm and a balance scale in her other hand. The Romans knew her as Ceres, their version of Demeter. Other representations of Virgo in the classical world included Astraea, the Star Maiden, and Justa, the Roman goddess of justice.

In China, the stars of Virgo were seen as part of a celestial Red Bird or Pheasant. Early Arabians saw these stars as part of an enormous Lion.

Description•Virgo lies in the general vicinity of the North Galactic Pole. As a result, when we look towards Virgo and other constellations in the area, we are looking at a region of space "above" the plane of the Milky Way Galaxy and therefore away from the spiral arm system of the Galaxy, which produces brilliant young stars, such as those that distinguish the constellations Orion and Scorpius. On the other hand, relatively low concentrations of interstellar dust in this direction favor observations of objects in intergalactic space, far beyond the confines of our Milky Way Galaxy.

The Virgo Cluster of galaxies is centered in the western part of the constellation, and, at 50 million light-years, it is the closest great cluster of galaxies to the Milky Way.

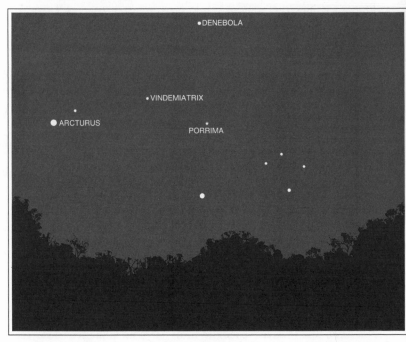

East	East-Southeast	Southeast

Porrima POUR-ih-mah Gamma, γ Virginis, page 124	2.8/3.7 magnitude blue-white F0 main sequence V 33 light-years −0.568 +0.008 −20	3.7 absolute magnitude 12h 42m −01° 27′ May 17 transit STR 743

Location•Porrima is a third-magnitude star found about 13 degrees to the south of Vindemiatrix.

Meaning•This star represents an ancient Latin goddess of prophecy.

Lore•In Babylon, Porrima was known as the Star of the Hero, while in China astronomers called this star the High Minister of State.

    Ulug-Beg noted that Porrima marked the Angle or Corner in a line of stars from Vindemiatrix through Delta, Gamma, Eta, and Beta in the constellation Virgo. This line marked a region of the sky known to Ulug-Beg as the Kennel.

History•In 1718 James Bradley found that Porrima was a double star. The orbit was calculated by John Herschel, and the components have a separation that varies between 6.0 and 0.4 arcseconds. The orbital period is about 171 years, and the next minimum separation will occur in the year 2007.

Description•The components of Porrima are virtually twin stars. Both are blue-white main-sequence objects. Porrima's distance is 33 light-years, near the standard for the calculation of absolute magnitudes. The combined magnitude is 2.8, and the individual stars each have an apparent magnitude of 3.7 and a luminosity three times that of the sun.

Stars of this type have surface temperatures of 7,400°K, with masses and diameters about 1.3 times greater than the sun's. These stars have estimated lifetimes in the order of magnitude of 4 billion years.

Nearby Features

M104: THE SOMBRERO GALAXY (12h 39m − 11° 31′)

M104, the lovely Sombrero Galaxy, is located 10 degrees to the south of Porrima. It derives its name from the striking appearance of dust lanes seen in long-exposure photographs.

This galaxy has an estimated distance of 45 million light-years, and with an apparent magnitude of 8, it may be glimpsed with small telescopes under excellent observing conditions. At such times, M104 may be detected with binoculars, but a telescope of at least 6 inches in diameter is needed in order to see a hint of the dust lanes.

See the Virgo constellation map on page 124.

QUASAR 3C273 (12h 29m + 2° 03′)

One of the brightest of the quasars is located 4.5 degrees to the northwest of Porrima and is designated as 3C273. This quasar has an apparent magnitude of nearly 13, and an analysis of its redshift leads to a distance estimate of about 2 billion light-years. Under good conditions, an experienced observer using a 6-inch telescope should be able to see this remarkable object.

See the Virgo constellation map on page 124.

Muphrid   March 31

East-Northeast   **East**   **East-Southeast**

Muphrid	2.7 magnitude	2.9 absolute
MUFF-rid	white G0	magnitude
Eta, η	subgiant IV	13h 55m $+18°$ 24′
Boötis, page 130	30 light-years	June 5 transit
	$-0.064 -0.363 -0$	STR 753

Location·Muphrid is a third-magnitude star located 5 degrees to the west of brilliant Arcturus. It is also 15 degrees northeast of Vindemiatrix.

Meaning·The name is derived from a description of the star by Ulug-Beg as Single Star of the Lancer.

Lore·In Arabian sky lore, Muphrid represented the companion of Arcturus, on whose lance the northern part of the canopy of heaven was supported.

Muphrid, along with its faint neighbors Nu and Tau, was known in China as an Officer at the Right Hand of the Emperor, with Arcturus representing the Emperor.

Traditional European portrayals of the constellation often show Muphrid marking the left knee of Boötes.

Description•Muphrid is a white subgiant with a luminosity six times that of the sun. Its surface temperature is about 5,800°K. Since it lies near the standard distance of 32.6 light-years used to calculate absolute magnitudes, the apparent and absolute magnitudes of Muphrid are nearly the same value.

Muphrid is a spectroscopic binary with a period of 494 days. There is a ninth-magnitude optical companion at a distance of 113 arcseconds.

Nearby Features

THE NORTH GALACTIC POLE (12h 49m +27° 24′)

The North Galactic Pole is located 16 degrees to the northwest of Muphrid. This pole marks the point on the celestial sphere that is 90 degrees from the plane of the Milky Way.

Since the plane of the Galaxy is where most of the Milky Way's dust is found, the regions around the Galactic Poles present a relatively clear view out towards intergalactic space. It is therefore no accident that most of the distant galaxies are seen in the general region of these Galactic "portholes."

Just a few degrees northeast of the North Galactic Pole, astronomers have discovered one of the largest known galaxy clusters — the Coma Cluster, estimated to be centered about 350 million light-years from the Milky Way and to contain thousands of individual galaxies.

See the Canes Venatici constellation map on page 118.

NGC 4889 (12h 59m +28° 07′)

NGC 4889 is a giant elliptical galaxy near the center of the Coma Cluster. It has an apparent magnitude of 11.4 and a redshift that, in addition to giving a clue to the galaxy's distance, indicates that it is rushing away from the Milky Way at a speed of about 6,500 kilometers per second.

NGC 4565 (12h 35m +26° 08′)

The lovely "edge-on" galaxy NGC 4565 is also located a few degrees from the North Galactic Pole, in this case to the west.

NGC 4565 is a spiral galaxy, oriented so that the plane of its disk and arm system is aimed nearly directly towards us. It is believed to be an outlying member of the Virgo Cluster of galaxies, at a distance of about 50 million light-years. Long-exposure photographs of NGC 4565 provide a beautiful illustration of the distribution of stars, luminous gas, and dust along the plane of a typical spiral galaxy.

With an apparent magnitude of 10.5, this galaxy may be glimpsed by experienced observers on clear, dark nights using telescopes having diameters of 8 inches or more.

See the Canes Venatici constellation map on page 118.

Boötes, *the Herdsman*

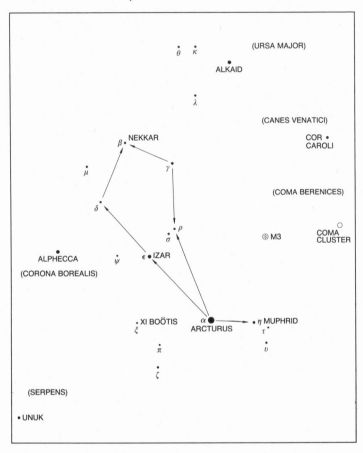

BRIGHT STARS OF BOÖTES

March 31	Muphrid	2.7	white	subgiant
April 5	Arcturus	−0.1	yellow	giant
April 7	Izar	2.4	yellow	giant

bow-OH-tease
Boo
Boötis

Evening Season • April – September

Sky Track • The stars of Boötes rise in the northeast, pass the meridian overhead, and set in the northwest.

Lore•Boötes, Ursa Major, Orion, Canis Major, the Hyades, and the Pleiades were the only star patterns referred to by both Homer and Hesiod. Since both wrote only about these stars, other traditional star groups had probably not yet been introduced into Greece at that time, in the ninth century B.C.

By the third century B.C., the astronomical writer Aratus mentioned a total of forty-five constellations, with most of these believed to have originated in Mesopotamia.

As is the case with other very bright stars, myths and legends originally dealing with Arcturus have been later applied to the entire constellation.

Arcturus and the rest of its constellation have long been described as the Bear Driver or Plowman. A legend tells that Boötes served to herd the Great Bear along its path around the northern sky. Another tale indicates that Boötes was a plowman, responsible for guiding the oxen that pulled the Plow, known in the United States as the Big Dipper.

An ancient Greek description of the constellation identifies Boötes as Icarius, a legendary Athenian who was taught by Bacchus the techniques of making wine.

Description•When the night is clear, the fainter stars to the north of Arcturus trace out the general shape of an ice cream cone or, as others see it, a kite. Once you become familiar with the brighter stars in a constellation, make up your own lines of connections between stars. In this way you can personalize a constellation so you may more easily recognize its fainter stars.

Nearby Features

XI BOÖTES (14h 51m + 19° 06′)

Xi Boötis is an attractive double star found 8 degrees due east of Arcturus. The components are both yellow main-sequence stars with magnitudes of 4.7 and 6.8, with a combined magnitude of 4.6 and a separation of 7 arcseconds. A 6-inch telescope with a magnification of 100× should be able to resolve the components of Xi Boötis, and also show a lovely field of stars in the immediate vicinity.

M3 (13h 41m + 28° 29′)

Eleven degrees to the northwest of Arcturus is located a globular cluster of stars known as M3. With binoculars this cluster looks like an out-of-focus star of magnitude 6.4, and an 8-inch telescope is needed to begin to resolve some of its individual stars. The distance of M3 is estimated to be about 35,000 light-years, and it probably contains over 500,000 stars.

April Stars

Arcturus

April 5

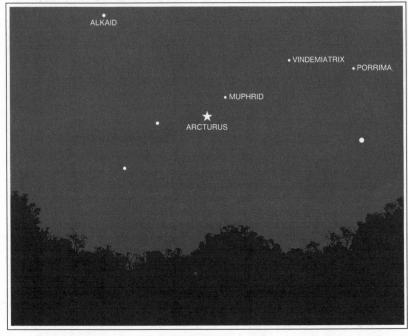

East-Northeast	East	East-Southeast
Arcturus arc-TO-rus Alpha, α Boötis, page 130	−0.1 magnitude yellow K2 giant III 36 light-years −1.098 −1.999 −5	−0.2 absolute magnitude 14h 16m +19° 11′ June 11 transit STR 771

Location • Arcturus is the most brilliant star in the evening skies of late spring and summer. It is found by following the curve of the Big Dipper's handle onward, about 30 degrees along an arc. This "arc to Arcturus" is a sure way to locate the star.

Meaning•The name has a Greek origin and means "the bear guard." Arcturus has appeared in literature since the time of Hesiod, about 800 B.C. Arcturus is one of the best-known stars and one of the few that has a proper name dating from ancient times.

Lore•The arrival of Arcturus in the eastern evening sky is a sure sign of spring. After a long, cold winter, the appearance of Arcturus is a welcome indication that the time of warm weather and flowers has returned.

This star was often associated with the Great Bear. "Bear Guard" refers to a traditional role for Arcturus as the Guardian or Keeper of the Bear, helping Ursa Major follow its assigned track around the northern sky.

In Arabia, Arcturus was called the Lofty Patriarch, or the Lofty Lance-bearer. This Lance was believed to be one of the supports of the canopy of heaven.

To the Shawnee Indians of Tennessee and South Carolina, Arcturus represented a great hunter known as White Hawk.

Arcturus has been shown in various places in the constellation figure of Boötes but is generally portrayed along the lower part of the Herdsman's coat.

History•Arcturus may have been the first star other than the sun to be seen during the daytime. This feat was accomplished in 1635 by Jean Baptiste Morin using a telescope equipped with setting circles, which enabled the instrument to be aimed at the star. Through Morin's telescope Arcturus could be seen, against a blue sky.

Description•Although Arcturus is one of the closest giant stars to the sun, it is a rather temporary visitor to our region of the Milky Way. It belongs to a category of stars called population II, first described by Walter Baade in 1944. These stars are usually found in the nucleus of a galaxy or in globular clusters. They are characterized by a relatively low content of elements other than hydrogen and helium. This suggests that they are remnants from the first generations of stars formed in the Galaxy, probably over 10 billion years ago.

Arcturus' steeply inclined orbital path happens to intersect the Galactic disk in the region of the sun. During most of its orbit around the center of the Milky Way, Arcturus is located far from the disk and spiral arms in a spherical region called the Galactic halo, a spherical volume of space that surrounds a spiral galaxy's core and disk.

Arcturus is also known as a high-velocity star because its highly inclined orbit results in a substantial motion relative to the sun and its other spiral-arm neighbors. Arcturus is slicing through the Milky Way's disk at a rate of about 150 kilometers per second, relative to the sun. Its proper motion is 2.3 arcseconds per year in the direction of the star Spica. Over a seventy-year period, Arcturus moves across the sky about one-tenth the apparent diameter of the full moon.

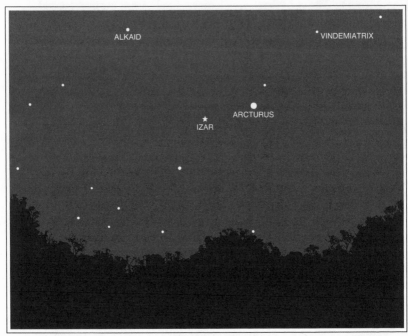

Northeast **East-Northeast** **East**

Izar	2.4/2.7 magnitude	−0.0 absolute
EYE-zar	yellow K1	magnitude
Epsilon, ε	giant III	14h 45m +27° 04′
Boötis, page 130	110 light-years	June 18 transit
	−0.051 +0.018 −23	STR 779

Location•Izar is found 10 degrees northeast of Arcturus.

Meaning•The name Izar is derived from a description, the Belt of Boötes.

History•Izar was first called Pulcherrima by Wilhelm von Struve, who discovered that this star is a double. Struve was one of the first astronomers to give extensive attention to the measurement of double stars. In 1829, he resolved the separation between Izar and its companion and noted the lovely color contrast of these stars.

Description•Izar is a yellow giant with a magnitude of 2.7 and a luminosity 80 times that of the sun. Its surface temperature is 4,400°K. Neutral metals present the most intense absorption lines seen in the spectrum of Izar. The star's proper motion is 0.051 arcsecond per year towards the west, and 0.018 arcsecond per year towards the north. The companion is a blue-white main-sequence star having a magnitude of 5.1, and the blended light of both stars is of apparent magnitude 2.4. The angular separation between Izar and its companion is about 3.0 arcseconds, and magnifications of over 200× may be needed to resolve the pair.

Izar and its companion have the same space velocity and therefore probably form a true binary pair. Since the separation of these stars has not changed greatly since Struve's observations, the orbital period probably is equal to several thousands of years.

The relative apparent distance between members of a double-star pair are described in terms of angular separation. The units of angular separation are arcseconds or arcminutes, measured across the celestial sphere.

Eta Draconis April 8

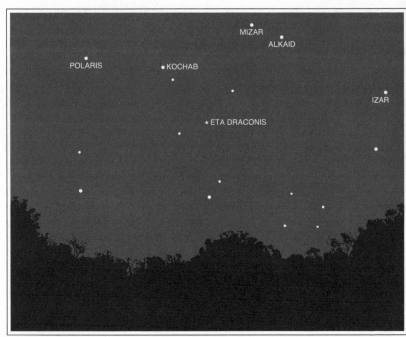

North	North-Northeast	Northeast

Eta Draconis A-tah drah-KOE- niss Eta, η Draconis, page 138	2.7 magnitude yellow-white G8 giant III 65 light-years −0.024 +0.059 −14	1.3 absolute magnitude 16h 24m +61° 31′ July 13 transit STR 785

Location•Eta Draconis is a third-magnitude star of the constellation Draco. Follow a line from Phecda through Megrez and extend it about 30 degrees to locate Eta.

History•Eta Draconis and its neighbor, Zeta, were sometimes known in Arabia as the Two Wolves, Black Bulls, or Ravens. Descriptions such as these were often applied to stars in order to render more familiar these glorious sparks in a realm so far beyond human understanding.

Description•Eta Draconis is a yellow-white giant star with an apparent magnitude of 2.7 and an estimated surface temperature of 4,700°K. The luminosity is about 25 times greater than that of the sun, and the distance to this star from the earth is 65 light-years.

The spectra of stars such as Eta Draconis contain prominent absorption lines characteristic of metals. The first proper motion value of 0.024 shown in the above data table tells us that Eta Draconis is moving towards the west (decreasing right ascension) at the rate of 0.024 arcsecond per year. The value +0.059 indicates this star moves on the celestial sphere towards the north (increasing declination) at the rate of 0.059 per year.

Eta's data table shows that the star's radial velocity is −14. This means that the distance between the sun and Eta Draconis is decreasing by 14 kilometers every second.

Draco, *the Dragon*

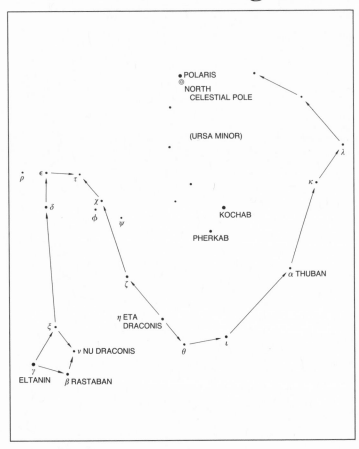

BRIGHT STARS OF DRACO

April 8	Eta Draconis	2.7	yellow-white	giant
May 3	Rastaban	2.8	white	giant
May 13	Eltanin	2.2	yellow	giant

DRAY-ko
Dra
Draconis

Evening Season•Draco is a circumpolar constellation, and its stars are visible during nights throughout the year. However, the constellation is most easily seen in the evening from April through October.

Sky Track•Northern circumpolar

Lore•In ancient Egypt, some of the stars of Draco were pictured as part of a falcon-headed man.

A Babylonian myth associates Draco with the dragon god Tiamat, who was subdued by the sun god represented as a kneeling figure just to the south of the creature's head. This "kneeler" was later incorporated into Greek sky lore as a representation of Hercules.

Greek legends sometimes saw Draco as the terrible dragon that was slain by Cadmus, brother of Europa and first king of Thebes. When the dragon was dead, Athena appeared and told Cadmus to sow the ground with the creature's teeth. From these seeds, a host of armed men grew, several of whom helped Cadmus found his new city-state.

Draco marked the Palace of the Heavenly Emperor to astronomers in China.

To Hindu astronomers, Draco represented an alligator or a porpoise.

Description•Draco winds around the northern sky, half enclosing the Little Dipper. Draco contains three stars brighter than magnitude 3.0, but most of the Dragon's figure is outlined by third-, fourth-, and fifth-magnitude stars.

Two of the brightest stars, Eltanin and Rastaban, mark the Dragon's head, while fourth-magnitude Thuban served as the pole star around 3000 B.C.

Giena   April 15

Southeast   **South-Southeast**   **South**

Giena	2.6 magnitude	−2.0 absolute
GEE-nah	blue-white B8	magnitude
Gamma, γ	giant III	12h 16m −17° 33′
Corvi, page 142	300 light-years	May 11 transit
	−0.163 +0.018 −4	STR 814

Location•Giena is in the small southerly constellation of Corvus, the Crow. It is a third-magnitude star located 15 degrees to the south of Porrima.

Meaning•The name Giena is derived from Ulug-Beg's description of this star as the Right Wing of the Raven.

Lore•The bird whose right wing is represented by Giena is variously described as a Crow or as a Raven.

Description•Giena is a blue-white giant star estimated to be 300 light-years from earth. It has a surface temperature of 10,100°K and an absolute magnitude of −2.0. Giena is about 500 times brighter than the sun.

If it were seen from earth at the distance of Sirius, Giena would shine with a brilliance about equal to Venus at its maximum brightness, 15 times brighter than Sirius.

Nearby Feature

DELTA CORVI (12h 30m − 16° 31′)

The star Delta Corvi, located 3.5 degrees to the west of Giena, is a wide double star having a separation of 24 arcseconds. The components have magnitudes of 3.0 and 8.0, with colors of blue-white and yellow. This pair is visible in small telescopes with a magnification above 100×.

See the following constellation map.

Corvus, *the Crow*

PORRIMA

(VIRGO)

SPICA

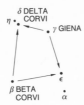

δ DELTA CORVI
η
γ GIENA
ϵ
β BETA CORVI
α

BRIGHT STARS OF CORVUS

April 15	Giena	2.6	blue-white	giant
May 16	Beta Corvi	2.7	yellow-white	giant

CORE-vuss
Cvr
Corvi

Evening Season•April – June

Sky Track•Corvus follows a southerly sky track, rising in the southeast and setting towards the southwest.

Lore•Although it covers a small area and has few bright stars, Corvus was a rather famous constellation in classical times. One well-known myth tells that Corvus the Crow was a cupbearer for Apollo, god of the sun. One day, the bird lingered at a tasty fig tree rather than speeding back to the heavens with a cup of water for the thirsty

Apollo, who was driving the solar chariot across the sky. When at last the Crow had eaten his fill of figs and returned to Apollo with the cup of water, he also carried a water snake in his beak. The Crow claimed that the snake had hindered him and that much time had been required to subdue the serpent. Apollo had no patience with this tale and banished the Crow to a place in the heavens next to Hydra, the water snake. The faint constellation Crater, the Cup, lies just to the west of Corvus, representing the once exalted position of the Crow.

The legend of a bird and a water snake predates Greece and probably originated in Mesopotamia. In ancient Israel, the stars of Corvus were also identified with a Raven. To astronomers in China, Corvus marked the Imperial Chariot, riding on the Wind.

An early Arabian celestial legend defined the stars of Corvus as the Throne of the Unarmed One. Spica, in the constellation Virgo, was known as the Unarmed due to its rather isolated location, without any bright stellar attendants.

Description· Although it is marked only by third-magnitude stars, the compact figure of Corvus is an interesting sight in an otherwise sparse portion of the sky.

Alphecca   April 19

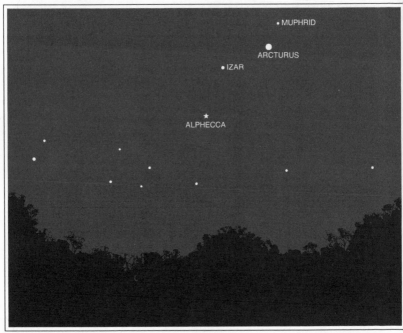

Northeast   **East-Northeast**   **East**

Alphecca	2.2 magnitude	0.5 absolute
al-FECK-ah	blue-white A0	magnitude
Alpha, α	main sequence V	15h 35m $+26°$ 43′
Coronae Borealis,	75 light-years	July 1 transit
page 146	$+0.120 -0.091 +2$	STR 829

Location•Second-magnitude Alphecca is the brightest star in the constellation Corona Borealis, the Northern Crown. Alphecca is located 18 degrees northeast of Arcturus and 10 degrees to the east of Izar.

Meaning•The name Alphecca originated with a description of Corona as the Broken One. This meant that the constellation figure is shown by a semicircle rather than a full circle of stars.

Lore•Other names for this star have been Gemma, the Gem, and the Pearl of the Crown. Alphecca's appearance in the asterism also resembles a diamond on a ring.

Alphecca marked the knot in a ribbon that tied together the leaves and flowers seen forming early representations of the Northern Crown.

Description· Alphecca is a blue-white main-sequence star 75 light-years from earth. It has a luminosity of 52 suns and a surface temperature of 9,900°K.

The mass of Alphecca is 2.5 times that of the sun. This value may be accurately calculated since Alphecca is a member of a binary system whose orbital characteristics and distance are well known. Alphecca is estimated to have a main-sequence life expectancy in the order of about 600 million years.

Alphecca is a spectroscopic binary whose orbit has been determined. The components of the pair have an orbital period of 17.36 days, and as they orbit, the stars partially eclipse each other, causing a brightness variation of about 0.1 magnitude.

The primary and secondary components are separated by about 28 million kilometers, about one-half the distance between the planet Mercury and the sun.

Corona Borealis, *the Northern Crown*

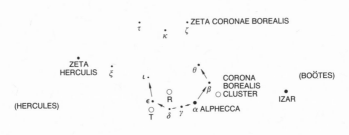

BRIGHT STAR OF CORONA BOREALIS

April 19	Alphecca	2.2	blue-white	main sequence

coe-ROW-nah BORE-ee-ALICE
CBr
Coronae Borealis

Evening Season•Corona is a feature of the evening sky from April through September.

Sky Track•Corona Borealis follows a northern sky track. It rises in the northeast, passes overhead, and sets towards the northwest.

Lore•This constellation was best known in classical antiquity as the Crown of Ariadne, daughter of King Minos of Crete. It was she who assisted Theseus in defeating the Minotaur and escaping from the labyrinth. Theseus took Ariadne with him, promising to marry her when they reached Athens. However, their ship made a stop at the island of Naxos, and Ariadne was abandoned there by Theseus. She later married the king of Naxos and her crown is said to be represented by the stars of Corona Borealis.

The stars of Corona were similarly known to the ancient Hebrews as a celestial Crown. In China, this group of stars was described as the Cord.

The Shawnee Indians knew these stars as the Celestial Sisters. The most beautiful of them, Alphecca, represented the wife of the great hunter, White Hawk. White Hawk was marked by the star Arcturus.

Arabian astronomical legends associated Corona with a celestial Dish, while Australian aborigines saw this group of stars as a Boomerang.

Description•The Northern Cross is a beautiful semicircle of stars that features the star Alphecca as its "gem."

Corona contains several faint variable stars, such as R Coronae Borealis and T Coronae Borealis, which are of interest because of their unusual light changes. In addition, the constellation indicates the direction of the very distant Corona Cluster of galaxies.

Nearby Features

R CORONAE BOREALIS (15h 47m +28° 19′)

In 1795 an English astronomer, Edward Pigott, discovered the irregular variable star now known as R Coronae Borealis, located 3 degrees of arc to the northeast of Alphecca. R Coronae Borealis usually shines with a magnitude of about 6, but suddenly and unpredictably it drops to a magnitude of between 7 and 14. One theory suggests that R Coronae Borealis is the hydrogen-depleted remnant of a very old star that, from time to time, ejects clouds of carbon, which for a year or so partly hide the star and cause it to appear considerably fainter.

T CORONAE BOREALIS (15h 57m +26° 04′)

Five degrees west of Alphecca is T Coronae Borealis, also called the Blaze Star. This is the brightest example of a class of stars known as recurrent novae.

Recurrent novae are believed to be relatively short period versions of ordinary novae. Novae are seen as the result of sudden, brilliant increases in the light of the denser component of certain very close binary systems that also contain a giant star. The components are so near each other that the denser member of the system gradually pulls gas away from its huge neighbor. After a period of perhaps many centuries, the accumulated mass of the smaller star becomes such a burden that a critical point is reached, a sudden burst of thermonuclear activity produces a brilliant flash of light, and the excess gas is blown away. Relieved of its extra mass, the nova star settles down and the gas transfer process begins again, eventually leading to another explosive episode. By contrast, supernovae are far more violent events that destroy extremely massive stars at the end of their evolution.

Normally a very faint, tenth-magnitude object, T Coronae Borealis both in 1866 and 1946 suddenly flared in brightness to the second magnitude and was easily visible to the naked eye.

ZETA CORONAE BOREALIS (15h 39m +36° 38′)

Zeta Coronae Borealis is a double star located 9 degrees north of Alphecca. Its components have magnitudes of 5.1 and 6.0, with a separation of 6.4 arcseconds. The primary appears yellow and the companion is blue-white; they may be observed with a small telescope at 100×.

CORONA BOREALIS CLUSTER OF GALAXIES

The Corona Borealis Cluster of galaxies is centered at a point in the sky about 2.5 degrees to the west of Alphecca. This rich and remote cluster is believed to be similar in size to the Virgo Cluster of galaxies. However, due to its distance, estimated to be between 1 and 3 billion light-years, the Corona Borealis Cluster covers only a small patch of sky, about equal to the apparent size of the full moon.

During the 1930s, galaxies in this cluster provided important information which helped establish the concept of an expanding universe. In the previous decade Albert Einstein, Wilhelm de Sitter, George Lemaitre, and Aleksandr Friedman developed theories on an expanding universe to explain why the universe did not collapse due to its own gravitational attraction. Also, astronomers Vesto Silpher, Edwin Hubble, and Milton Humason observed large redshifts in the spectra of distant galaxies. This indicated that these objects were moving away from us at high rates of speed. In 1929 Hubble described a direct relationship between the amount of a galaxy's redshift and its distance from the Milky Way. Hubble's "law" implied that the universe was indeed expanding. By 1936 Humason had discovered redshifts of several galaxies in the Corona cluster which indicated radial velocities of +21,600 kilometers per second. These observations offered very firm evidence for the expansion of the universe. During the 1950s the Big Bang theory was introduced to account for the origin of the universe and its expansion which began about ten to twenty billion years ago. The discovery of cosmic background radiation in 1964 provided evidence of the light flash from the primordial fireball of the Big Bang. The light was now stretched into radio wavelengths. During 1967 William Fowler, Fred Hoyle, and Robert Wagoner outlined the physical events of the Big Bang, and in 1979 Steven Weinberg won the Nobel Prize for his study of the event's first moments. The "inflationary universe" theory was proposed by Alan Guth in 1979 to enhance understanding of the origins and expansion of the universe.

The Big Bang theory has stimulated much scientific and philosophical thought. A rather interesting parallel might be seen between the concept of an expanding universe and several interpretations of a Hebrew word, "raqia," usually translated in Scripture as either firmament or expanse. In 1762 English lexicographer John Parkhurst noted that *raqia:* "expresses motion of different parts of the same thing, at the same time, one part the one way and the other, the other way, with force." He defined *raqia* as: "an expansion the celestial fluid or heavens in a state of expansion." In 1821, John Reid of Scotland gave this definition: "Expansion, the heavens, from their being stretched forth." In 1983, Old Testament scholar Roland K. Harrison considered the varied renderings of *raqia:* "This curious divergence of meaning is matched by the difficulty in translating the original Hebrew term raqi(a). It is a cognate form of the verb rq, '(to) spread out,' or '(to) beat out,' the former usage referring to the expanse of the heavens at creation, and the latter to the beating out of metal into thin plates or sheets. . . . The word firmament (raqia) occurs only in the Old Testament, and always within the context of creation."

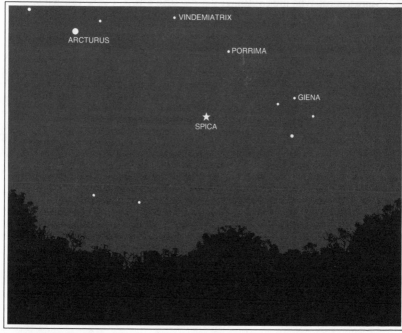

East-Southeast	Southeast	South-Southeast

| Spica
SPY-ka
Alpha, α
Virginis, page 124 | 1.0/3.1 magnitude
blue-white B1
giant-subgiant III –
IV
250 light-years
$-0.043 - 0.033 + 1$ | -3.4 absolute
magnitude
13h 25m $-11°$ 10′
May 29 transit
STR 832 |

Location•Spica is a first-magnitude star in the constellation Virgo. During his many years as lecturer at the Hayden Planetarium, Henry M. Neely taught this helpful phrase: "Make an arc to Arcturus and speed on to Spica." Remember these words as you follow the curve of the Big Dipper's handle to Arcturus and continue to Spica for a sure identification of this star.

Meaning•*Spica* is a Latin word meaning "ear of grain." The Romans used this name because this star marked the sheaf of wheat held by the goddess Ceres, depicted in the constellation figure of Virgo.

Lore•Because it is bright, and very near to the ecliptic, Spica served as an important place marker in the zodiac during ancient times. In

Babylon, the star was known as the Virgin's Belt. Chinese astronomers knew Spica as the Special Star of Springtime, and in India the descriptions Bright One, Lamp, and Pearl were used. Spica was called the Unarmed in Arabian sky lore. It, along with Arcturus, was seen as a post that supported the canopy of heaven. The term *unarmed* indicated Spica's lack of close, bright neighboring stars.

Description•Spica is a multiple-star system. The primary star has a blended apparent magnitude of 1.0. It is the fifteenth-brightest star in the night sky, and the brightest component has an apparent magnitude of 3.1, a surface temperature of about 25,500°K, and luminosity 2,000 times greater than the sun. Its estimated main-sequence life expectancy is in the order of magnitude of 50 million years.

Occultation studies indicate apparent magnitude of 4.5 and 7.5 for the other components on the primary set of stars. The secondary component of Spica is a blue-white main-sequence star of spectral class B2. It was discovered through spectral analysis by H. C. Vogel in 1890. In addition, a twelfth-magnitude visual companion lies at a distance of 148 arcseconds.

Nearby Feature

W VIRGINIS (13h 23m −03° 07′)

A faint, tenth-magnitude variable star known as W Virginis is located 8 degrees directly to the north of Spica. This was the first discovered example of a class of variable stars known as population II Cepheids. The pattern of light variation of these stars is somewhat similar to classical Cepheids, and a distinction was not recognized until 1950.

W Virginis stars are generally found in globular clusters and elsewhere in the Milky Way Galaxy's halo. As is the case with classical Cepheids, the period of light variation is an indication of the star's absolute magnitude, therefore the period also provides a clue to the star's distance. W Virginis stars are about 1.5 magnitudes fainter than classical Cepheids in the Galactic disk that have the same period of light variation.

W Virginis is a supergiant star. Its faint apparent magnitude is a result of the star's great distance of 11,000 light-years from earth. W Virginis is located in the region of the Milky Way's halo, far from the plane of the Galactic disk.

See the Virgo constellation map, page 124.

The Bright Stars of Spring

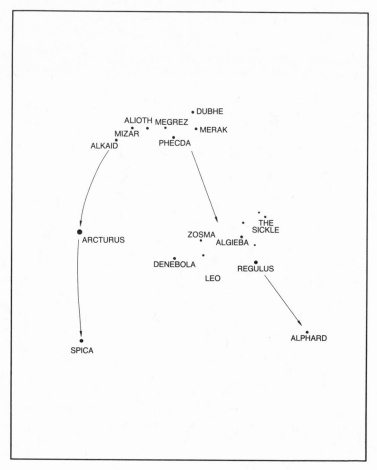

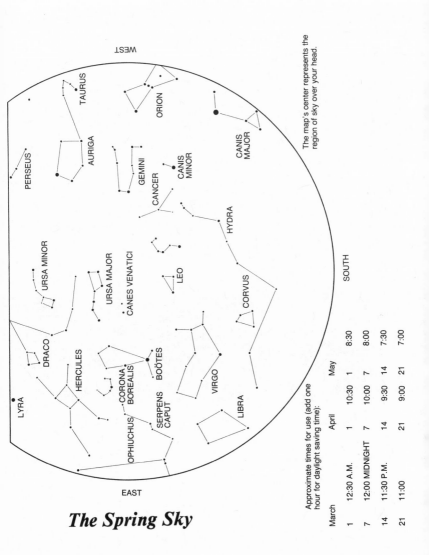

The map's center represents the region of sky over your head.

WEST

TAURUS
ORION
PERSEUS
AURIGA
GEMINI
CANIS MINOR
CANCER
CANIS MAJOR
HYDRA
URSA MINOR
URSA MAJOR
CANES VENATICI
LEO
CORVUS
DRACO
LYRA
HERCULES
CORONA BOREALIS
BOÖTES
OPHIUCHUS
SERPENS CAPUT
VIRGO
LIBRA

SOUTH

EAST

Approximate times for use (add one hour for daylight saving time):

March		April		May	
1	12:30 A.M.	1	10:30	1	8:30
7	12:00 MIDNIGHT	7	10:00	7	8:00
14	11:30 P.M.	14	9:30	14	7:30
21	11:00	21	9:00	21	7:00

The Spring Sky

Rastaban   May 3

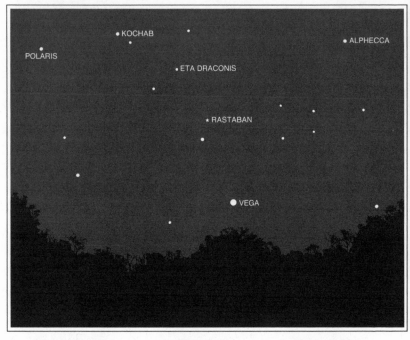

North-Northeast	Northeast	East-Northeast

Rastaban	2.8 magnitude	−3.5 absolute
RAS-tah-ban	white G2	magnitude
Beta, β	giant Ib–II	17h 30m +52° 18′
Draconis, page 138	400 light-years	July 30 transit
	−0.022 +0.013 −20	STR 882

Location•Rastaban is a third-magnitude star located in the sinuous constellation of Draco, which winds around the Little Dipper. On spring evenings, Rastaban is found about 50 degrees below Alkaid, the star at the end of the Big Dipper's handle.

Meaning•The name Rastaban comes from Head of the Snake, or Dragon's Head, this star's position in the constellation figure of Draco.

Lore•Rastaban and its second-magnitude neighbor, Eltanin, were known as the Dragon's Eyes in Arabian sky lore.

History•In 1889, the American astronomer S. W. Burnham discovered a twelfth-magnitude companion to Rastaban at a distance of 4.2 arcseconds.

Description•Rastaban is a white giant star that shines in our night sky with an apparent magnitude of 2.8. Its surface temperature is about 5,400°K.

The parallax measurement of Rastaban is too small to be relied upon as a fully accurate indication of this star's distance. As a result, spectral analysis is employed to estimate Rastaban's distance. The study of Rastaban's spectral lines results in a luminosity classification of Ib–II, which suggests an absolute magnitude of −3.5. Comparison of this number with the apparent magnitude provides a distance estimate that is then averaged with the distance indicated by the star's parallax. The resulting estimate of Rastaban's distance is about 400 light-years.

The specific appearance of the star's spectral lines implies a luminosity of nearly 3,000 times that of the sun for this star in the constellation Draco.

Zeta Herculis   May 4

Northeast	East-Northeast	East

Zeta Herculis ZAY-tah HER-kyou-lis Zeta, ζ Herculis, page 158	2.9 magnitude white G0 subgiant IV 32 light-years −0.471 +0.394 −70	2.9 absolute magnitude 16h 41m +31° 36′ July 18 transit STR 883

Location•Zeta Herculis is located 17 degrees to the northeast of Alphecca. Third-magnitude Zeta marks the southwestern corner of an asterism known as the Keystone of Hercules.

Lore•The Keystone is a faint but famous group of stars that represents the body of Hercules, mightiest hero of ancient Greek mythology.

History•In 1782 William Herschel found Zeta Herculis to have a close sixth-magnitude companion.

Description• The Zeta Herculis binary-star system is located 32 light-years from the earth. This is nearly equal to the standard distance of 32.6 light-years used to calculate absolute magnitude. As a result, the apparent magnitudes of the component stars of the Zeta system, 2.9 for the primary star and 5.5 for the secondary, are almost equal to their absolute magnitudes.

The companion to Zeta Herculis is a yellow main-sequence star at a distance of 1.5 arcseconds from the primary. This pair has an orbital period of 34.4 years. A solution of the mathematical equations that describe this binary system indicates a mass about 1.3 times that of the sun for zeta and 0.7 times that of the sun for the secondary star. The respective luminosities are 5.2 and 0.5 times that of the sun.

Nearby Features

M13: GLOBULAR CLUSTER (16h 41m +36° 30′)

Along the western side of the Hercules Keystone 4.5 degrees north of Zeta Hercules lies the magnificent globular cluster M13, brightest example of its type seen in northern skies. This is an ancient spherical assembly of possibly one million stars, located about 25,000 light-years from the earth and having a diameter of about 200 light-years. The cluster was discovered by Edmund Halley in 1714. To the naked eye, M13, also called the Hercules Cluster, looks somewhat like a faint, hazy star of the sixth magnitude, just on the threshold of visibility. Binoculars make the object easier to locate, and an 8-inch telescope begins to reveal individual stars along the cluster's fringe. As seen through telescopes with a diameter of 12 inches or more, M13 is an extraordinary sight. With the cluster resolved into its individual stars, it resembles a vision of diamond dust sprinkled onto a black-velvet cushion.

About 100 globular clusters have been discovered. They appear to be remnants of the first stages in the Galaxy's formation, which is believed to have taken place at least 10 billion years ago.

During 1917, the young astronomer Harlow Shapley of Mount Wilson Observatory surveyed the distribution of ninety-three globular clusters seen in various parts of the sky. Observations of Cepheid variable stars within these clusters provided Shapley with information about the clusters' distances. He assumed that the globular clusters had a spherical distribution in space centered at the "middle" of the Milky Way. This array of clusters provided an estimate of the Galaxy's size as well as the distance of the sun from its center.

M92: GLOBULAR CLUSTER (17h 17m +43° 10′)

Another well-known globular star cluster, M92, is located 9 degrees to the northeast of M13, also in the constellation Hercules. It is estimated to be about 28,000 light-years from earth.

See the following constellation map.

Hercules, *the Hero*

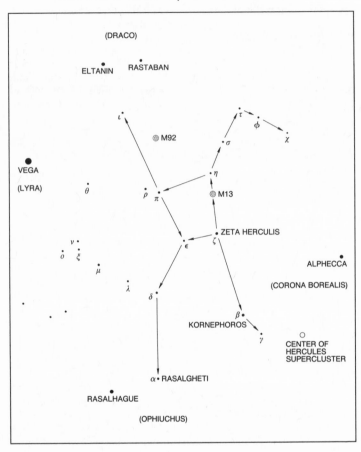

BRIGHT STARS OF HERCULES

May 4	Zeta Herculis	2.8	white	subgiant
May 8	Kornephoros	2.8	yellow-white	giant

HER-kyou-leez
Her
Herculis

Evening Season • May – November

Sky Track • The stars of Hercules rise in the northeast, pass overhead, and set towards the northwest.

Lore•This constellation was known to early Greeks as the Kneeling One, long before the name Hercules was adopted. The figure was sometimes described as having one of its feet placed triumphantly upon the head of Draco, the Dragon. This constellation originated in the Middle East, where it represented Gilgamesh, the epic hero of Sumerian and Babylonian mythology. Gilgamesh was portrayed kneeling in victory over the defeated dragon Tiamat, whose stars are known to us as the constellation Draco. A fourth-magnitude star, Tau Herculis, marks the hero's bended knee.

The name Hercules was applied to these stars by the Greeks sometime around the sixth century B.C. Edith Hamilton describes the character of Hercules in her *Mythology.*

> Hercules was the strongest man on earth and he had the supreme self-confidence magnificent physical strength gives. . . . Whenever he fought with anyone the issue was certain beforehand. He could be overcome only by a supernatural force. . . . His intellect was not strong. His emotions were. . . . Nevertheless he had true greatness. Not because he had complete courage based upon overwhelming strength . . . but because, by his sorrow for wrongdoing and his willingness to do anything to expiate it, he showed greatness of soul.

Hercules had to endure Twelve Labors as an atonement for the murder of his family in a fit of madness, induced by the vengeance of Hera. The Queen of Olympus was resentful because Hercules had been born from the union of Zeus and the mortal woman Alcmene.

Description•The central feature of Hercules is a keystone-shaped asterism of third- and fourth-magnitude stars located halfway between Vega and Alphecca.

During late Sumerian and early Babylonian times, about 4,000 years ago, the stars of Hercules were (due to precession) about 25 degrees closer to the North Celestial Pole than they are now. Observers of that period, from latitudes of the Tigris-Euphrates, had their best view of the Kneeler high in the sky as they faced north. From this perspective, the figure's head was near the zenith, with feet below, towards the northern horizon.

Nearby Feature•The double stars Delta Herculis, which has a separation of 10 arcseconds, and Rasalgeti, with a separation of 4.5 arcseconds, are also of interest, seen with eight-inch and larger telescopes.

East	East-Southeast	Southeast
Unuk YOU-nuk Alpha, α Serpentis, page 162	2.7 magnitude yellow K2 giant III 60 light-years +0.136 +0.044 +3	+1.3 absolute magnitude 15h 44m +6° 26′ July 3 transit STR 898

Location•Third-magnitude Unuk is in the constellation Serpens and is 20 degrees to the south of Alphecca.

Meaning•The name Unuk comes from the description Neck of the Serpent.

Lore•The serpent was associated in Greek mythology with a symbol of the divine healer Asclepius, who became the god of medicine and appears in the sky as the constellation Ophiuchus.

History•A twelfth-magnitude optical companion to Unuk was discovered by John Herschel in 1836. The separation is now about 60 arcseconds.

Description•Unuk is a yellow giant 60 light-years from earth. Its surface temperature is about 4,200°K, and we see it shine in the sky with a magnitude of 2.7. The luminosity of Unuk is estimated to be about 25 times that of the sun.

Nearby Feature

M5: GLOBULAR CLUSTER (15h 18m +02° 11′)

The sixth-magnitude globular cluster M5 is located 7.5 degrees southwest of Unuk. It is one of the brightest of these clusters seen in northern skies. M5 is believed to be about 30,000 light-years from earth.

This cluster was discovered by Gottfried Kirch, director of the Berlin Observatory, in May of 1702. It was not resolved into its individual stars until 1791, when the feat was accomplished by William Herschel using one of his large reflecting telescopes.

Serpens, *the Serpent*

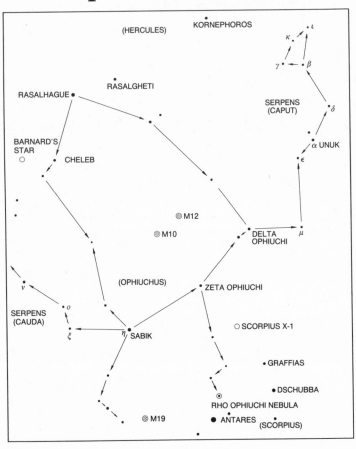

BRIGHT STAR OF SERPENS

May 7	Unuk	2.7	yellow	giant

SIR-pens
Ser
Serpentis

Evening Season•June – September

Sky Track•The Serpent rises around the eastern point of the horizon, passes the meridian halfway between the horizon and the zenith, and sets towards the west.

Lore•The head of the Serpent is located just south of Corona Borealis. The constellation winds through the sky towards the east, crosses the figure of Ophiuchus, and extends towards Aquila.

Serpens is officially listed as a separate constellation; however, it has traditionally been closely associated with Ophiuchus, the Serpent Holder. The Serpent's head, to the west of Ophiuchus, is known as Serpens Caput, while the tail, which is east of Ophiuchus, is called Serpens Cauda. The Serpent is a symbol of Ophiuchus' healing powers and is shown in his hands.

Serpens' profile is formed by a twisting line of third-, fourth-, and fifth-magnitude stars.

Kornephoros

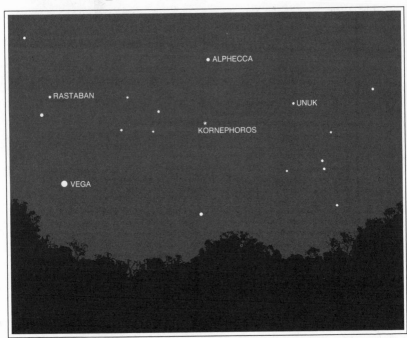

East-Northeast	East	East-Southeast

Kornephoros	2.8 magnitude	+0.7 absolute
kor-NEF-oh-ros	yellow-white G7	magnitude
Beta, β	giant III	16h 30m +21° 29′
Herculis, page 158	110 light-years	July 15 transit
	−0.099 −0.017 −26	STR 900

Location·Third-magnitude Kornephoros is 10 degrees south of Zeta Herculis and 20 degrees northeast of Unuk.

Meaning·This star name comes from the Greek for Club-bearer, a title for the constellation Hercules.

Lore·In the ancient descriptions of Hercules, the figure was shown with its head on the same side of the Keystone as Kornephoros. As a result of precession, this traditional portrayal causes Hercules to appear upside down as we see the constellation cross the meridian very high in the southern sky, nearly at the zenith.

Description·Kornephoros is a yellow-white giant star 110 light-years from the earth. Its apparent magnitude is 2.8, making this star 44 times more luminous than the sun. The surface temperature is about 4,850°K.

Kornephoros has a spectroscopic binary companion with a period of 411 days.

Nearby Feature

HERCULES SUPERCLUSTER OF GALAXIES (c. 16h + 18°)

It is believed that during the early stages in the development of the universe, galaxies formed in immense, stringy chains within pancake-shaped regions of space. We call these vast groups of galaxies superclusters. In a portion of sky about 5 degrees in diameter, centered 7 degrees to the southwest of Kornephoros, is located the Hercules Supercluster of galaxies.

An average of estimates of the distance to the center of this supercluster is about 500 million light-years. The system is made up of many individual clusters, some of which contain thousands of galaxies. It is the next closest aggregation similar to our own Local Supercluster of Galaxies, centered at the Virgo galaxy cluster. The Local Group of Galaxies, the small cluster that contains the Milky Way, belongs to this Local Supercluster.

See the Hercules constellation map, page 158.

It has been suggested that the superclusters contain 90 percent of all galaxies but occupy only 10 percent of the volume in the universe. Great voids between superclusters, in which few galaxies are found, have been discovered in the direction of the constellations Boötis and Cancer.

Eltanin

North-Northeast	Northeast	East-Northeast

Eltanin	2.2 magnitude	−0.6 absolute
el-TAY-nin	yellow K5	magnitude
Gamma, γ	giant III	17h 57m +51° 29′
Draconis, page 138	120 light-years	August 6 transit
	−0.013 −0.020 −28	STR 912

Location·Eltanin is a second-magnitude star 4 degrees of arc from Rastaban in the constellation Draco, the Dragon.

Meaning·The name was part of a description of the Dragon's head.

History·The eighteenth-century astronomer James Bradley carefully observed the position of Eltanin beginning on the night of December 3, 1725. By December 17, Bradley noted that Eltanin crossed the meridian slightly to the south of its position on December 3. By March of 1726, the star had reached a distance about 20 arcseconds to the south of its starting point, marking what would be its maximum deviation in this direction. During the following weeks, Eltanin was seen to begin moving slowly towards the north. It passed the first observed location, and during early September the maximum northern variation was reached. By December 1726, Eltanin was again seen at nearly the same spot, relative to the celestial coordinate system, as when Bradley had begun his observations the previous year.

Bradley could not explain these changes either in terms of parallax or atmospheric refraction. He therefore embarked on a program in which he studied nearly 200 stars in a similar way. By 1728 he realized that the shape of the path traced by a star depended on its distance from the ecliptic and therefore was probably associated with the earth's orbital motion around the sun.

In a moment of inspiration while taking a boat ride on the Thames, Bradley is said to have noted that the direction of a pennant flying from a mast relative to a boat, depends on the combination of wind direction and direction of the changing course of the boat. He made an association between the speed and direction of starlight and the motion of the earth as it travels on its orbit each year.

The approximate value of the speed of light had been determined in 1675 by Ole Rømer, and Bradley compared this speed to the velocity of earth's motion and was able to account for the variation, which he called aberration of starlight.

Description•Eltanin is a yellow giant at a distance from earth of about 120 light-years. The surface temperature is 3,800°K, and the luminosity is 145 times that of our sun.

This is a northern circumpolar star and, along with its neighbors in this part of the sky, is visible at all times of the night throughout the year.

Nearby Feature

NU DRACONIS (17h 32m + 55° 10′)

The attractive double star Nu Draconis is located in the "head" of Draco, 5 degrees from Eltanin and 3 degrees from Rastaban.

Nu Draconis consists of a pair of stars with a separation of 62.3 arcseconds. The components have nearly identical brightnesses, with magnitudes of 4.88 and 4.87. Both are blue-white main-sequence stars at a distance of 93 light-years from earth. This pair may be easily resolved with a small telescope, and binoculars may show the separation of these lovely stars.

Zubeneschamali May 15

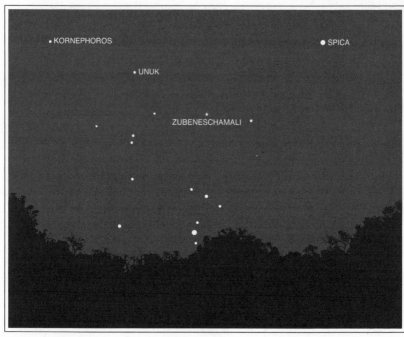

East-Southeast	Southeast	South-Southeast
Zubeneschamali zoo-BEN-ess-sha-MAY-lee Beta, β Librae, page 170	2.6 magnitude blue-white B8 main sequence V 130 light-years −0.098 −0.023 −35	−0.4 absolute magnitude 15h 17m −9° 23′ June 26 transit STR 932

Location•Zubeneschamali is a third-magnitude star found 17 degrees south of Unuk and 27 degrees to the east of Spica.

Meaning•The original form of the name meant the Northern Claw in Arabic.

Lore•Zubeneschamali was originally part of the constellation Scorpius and represented that creature's upper claw. In ancient Babylon, this star is said to have been called the Northern Light.

History•Zubeneschamali and its neighbor to the south, Zubenelgenubi, served ancient astronomers as markers of the moon's progress across the sky. Similar groups of stars, distributed along both sides of the ecliptic, were referred to throughout the Middle East, India, and China as lunar mansions or lunar stations.

Description•Zubeneschamali is a blue-white main-sequence star having an apparent magnitude of 2.6. Its distance from earth is estimated to be about 130 light-years. This star has a surface temperature of 10,100°K and shines with 120 times the sun's luminosity. The spectrum of this type of star is characterized by absorption lines of hydrogen and neutral helium. The prominent hydrogen lines of class B8 main-sequence stars, such as Zubeneschamali, are more intense and broader than those seen in B8 stars of higher categories of luminosity. As a result, the appearance of these lines helps identify this star as a member of the main sequence. In the spectrum of a class B8 supergiant, such as Rigel for example, these hydrogen lines are considerably narrower and less intense. Zubeneschamali is moving closer to the sun at the rate of 35 kilometers per second.

Main-sequence stars similar to Zubeneschamali are estimated to have masses about three times that of the sun and life expectancies in the order of about 250 million years.

Libra, *the Scales*

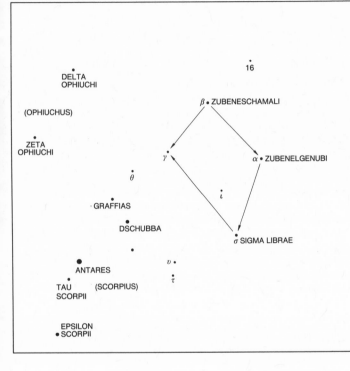

BRIGHT STARS OF LIBRA

| May 15 | Zubeneschamali | 2.6 | blue-white | main sequence |
| May 21 | Zubenelgenubi | 2.8 | blue-white | subgiant |

LEE-bra
Lib
Librae

Evening Season•May – June

Sky Track•Libra moves across the southern sky from southeast to southwest.

Lore•Libra was shown as a beam and scales on the Dendera planisphere (see Leo, the Lion, February 3), and it has been suggested that this stellar image originated in Egypt. Libra was also known as a scale or balance in Mesopotamia and Israel.

In Mesopotamia, the stars of Libra were also portrayed as a chariot yoke, an altar, incense burner, or as a lamp. At times this lamp was shown being held between the claws of Scorpius, the Scorpion. The names of the two brightest stars in Libra, Zubenelgenubi and Zubeneschamali, originated from their descriptions as the northern and southern claws. By the second century B.C., Libra was generally shown to represent the Claws of the Scorpion, although the scale's image had not been entirely forgotten in Greece.

The revival of Libra in the form of a scale is dated to the year 46 B.C., when the figure was introduced in Rome in the Julian calendar. During Roman times, Libra was described as Caesar Holding the Scales of Justice. However, by the Middle Ages, the Scales came to be associated with Virgo, the site of the sun at the time of the autumnal equinox, when the days and nights are of equal length.

Before the introduction of Western constellations, Libra was known in China as the Star of Longevity. In ancient India, the stars Zubenelgenubi and Zubeneschamali, as well as Sigma Librae, were called the Gateway, because the sun, moon, and planets, moving along the zodiac, passed between these stars.

Description•Although Libra has a rich mythological history, its brightest stars are only of the third magnitude.

Look for the "beam" of the Scales of Libra about 20 degrees to the east of Spica. This asterism is formed by the three brightest stars of the constellation, Zubeneschamali, Zubenelgenubi, and Sigma.

Beta Corvi   May 16

South-Southeast	South	South-Southwest
Beta Corvi BAY-tah CORE-vi Beta, β Corvi, page 142	2.7 magnitude yellow-white G5 giant II 95 light-years +0.001 −0.058 −8	+0.3 absolute magnitude 12h 34m −23° 24′ May 16 transit STR 934

Location•Beta Corvi is very similar in brightness to its third-magnitude neighbor, Giena, located 8 degrees to the northwest.

Lore•Beta is part of the figure of Corvus, the Crow. It also marked part of a Chinese asterism representing the Crosspiece of a Chariot. Beta and other stars in Corvus were part of the figure of the Hand, one of India's lunar mansions, markers of the moon's progress.

Description•This star is a yellow-white giant about 95 light-years from us. Beta Corvi has a surface temperature of 4,900°K and a luminosity of 65 suns. Its apparent magnitude is 2.7.

Both ionized and neutral metals present the most intense absorption lines in the spectrum of Beta Corvi. Ionized calcium shows especially strong lines. Beta Corvi's proper motion in right ascension is +0.001 arcsecond per year. The plus sign tells us that this is in the direction of increasing right ascension, so the motion is eastward. The star's proper motion in declination is −0.058 arcsecond per year. The minus sign indicates decreasing declination, or southward motion. Beta Corvi's radial velocity is −8 kilometers per second. Calculations, taking into account the star's proper motion, radial velocity, and distance from earth, show that Beta Corvi is moving through space at the rate of approximately 11 kilometers per second.

Although Beta Corvi and its neighbor Giena appear to be near each other and similar in brightness, you can see from their data tables that differences in distance, proper motion, and radial velocity are significant. Therefore, these stars are not related.

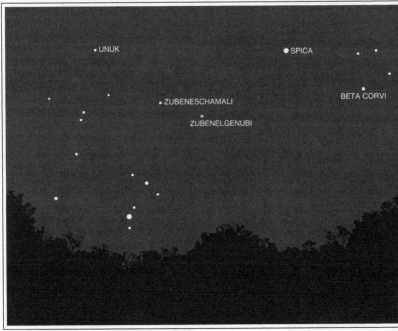

East-Southeast	Southeast	South-Southeast

Zubenelgenubi	2.8 magnitude	+1.2 absolute
zoo-BEN-el-je-	blue-white A3	magnitude
NEW-be	subgiant IV	14h 51m −16° 03′
Alpha², α^2	65 light-years	June 20 transit
Librae, page 170	−0.108 −0.071 −10	STR 952

Location•Third-magnitude Zubenelgenubi lies 9 degrees to the southwest of Zubeneschamali and is about one-half a degree north of the ecliptic. Spica is 21 degrees towards the west.

Meaning•Zubenelgenubi represented the Southern Claw of the ancient extended constellation of Scorpius. This star and its neighbor, Zubeneschamali, maintained their original identities when they were incorporated into the constellation Libra, the Scales.

Lore•Babylonian astronomers called Zubenelgenubi the Southern Light.

In India, Zubenelgenubi and its northeastern neighbor represented a celestial Gateway stretched across the zodiac, because the sun, moon, and planets passed between these stars on their journey along the zodiac.

Description•Zubenelgenubi is a blue-white subgiant of magnitude 2.8. It has a white companion star of magnitude 5.2, at a distance of 231 arcseconds. Both components are visible with the help of binoculars. The primary and secondary, also a subgiant, have nearly identical proper motions and are both about 65 light-years from earth. The brighter companion is designated Alpha^2 Librae, while the fifth-magnitude star is known as Alpha^1. These numbers were assigned on the basis of order in right ascension rather than of brightness.

Alpha^2 Librae, Zubenelgenubi, is about 25 times more luminous than the sun and has a surface temperature of 8,900°K.

Delta Ophiuchi   May 24

East	East-Southeast	Southeast

Delta Ophiuchi	2.7 magnitude	+0.4 absolute
DEL-tah oh-fee-	yellow-orange M0.5	magnitude
YOU-kye	giant III	16h 14m −3° 42′
Delta, δ	95 light-years	July 11 transit
Ophiuchi, page 178	−0.048 −0.145 −20	STR 965

Location•Delta Ophiuchi is located about 12 degrees southeast of Unuk and 15 degrees northeast of Zubeneschamali. It has a magnitude of about 3.0.

Lore•Delta was sometimes known as Yed Prior, the First Star in the Hand. It, along with Epsilon, represented the left hand of Ophiuchus, grasping the Serpent.

Description•Delta Ophiuchi is a yellow-orange giant. Measurements of its parallax provide a distance estimate of 95 light-years. Delta has a luminosity almost 60 times greater than the sun and a surface temperature of 3,200°K. Its apparent magnitude is 2.7.

Delta Ophiuchi is often called a "red" giant and most of its energy is being emitted into space in the form of infrared electromagnetic radiation. Neutral metals are the source of the most intense absorption lines in this star's spectrum. Lines characteristic of molecules of titanium oxide are also present.

Every year Delta's proper motion causes a decrease in its right ascension by 0.048 arcsecond. During this period the declination decreases by 0.145 arcsecond. The star's total proper motion is 0.153 arcsecond. At this rate Delta Ophiuchi would require about 12,000 years to cover a distance on the celestial sphere equal to the diameter of the full moon's disk.

Ophiuchus, *the Serpent Holder*

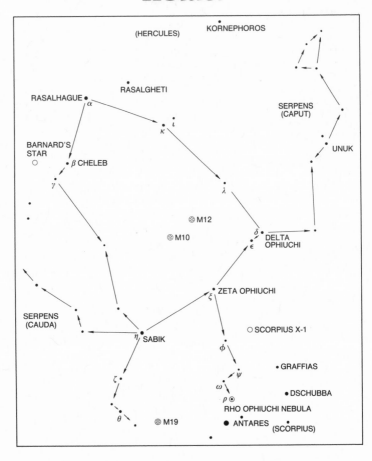

BRIGHT STARS OF OPHIUCHUS

May 24	Delta Ophiuchi	2.7	yellow-orange	giant
May 30	Rasalhague	2.1	blue-white	giant
June 7	Cheleb	2.8	yellow	giant
June 8	Zeta Ophiuchi	2.6	blue-white	main sequence
June 26	Sabik	2.4	blue-white	main sequence

OFF-ih-YOU-kus
Oph
Ophiuchi

Season•June – September

Sky Track•The stars in Ophiuchus are centered on the celestial equator, and as a result they rise in the east and set towards the west. Ophiuchus crosses the meridian about halfway between the horizon and the zenith.

Lore•Ophiuchus occupies the portion of sky between the constellation Hercules and Scorpius, on the western edge of the band of the Milky Way.

The Serpent held in the hands of Ophiuchus is listed as a separate constellation, although the two are linked in mythology.

The name Ophiuchus was used by early Greek astronomers, and the figure was most often associated with a mythical healer called Asclepius, said to have been a son of Apollo. As a boy he was sent to the wise centaur Chiron, who taught Asclepius about the curative use of herbs as well as other forms of medicine. In time, Asclepius became unsurpassed in the skills of healing and served as physician of the Argonauts.

The myths also tell that Asclepius-Ophiuchus became so skilled that he was able to revive the dead. At that point he provoked Pluto, who feared for the continuance of his underworld kingdom. Pluto induced Zeus to kill Asclepius with a thunderbolt, and to place him in the sky as a constellation.

His memory was venerated, and the serpent came to symbolize his wisdom and healing powers. The great physician Hippocrates was said to have been a direct descendant of Asclepius.

Description•The stars that outline the figure of Ophiuchus are of moderate brightness, and as a result the constellation is not nearly as eye-catching as are its southern neighbors, Scorpius and Sagittarius.

Although Ophiuchus is not one of the ancient zodiacal constellations, the sun is within the borders of Ophiuchus for sixteen of the twenty-five days required for it to move from Libra to Sagittarius. During this period, the sun is within the boundaries of Scorpius for only nine days.

Ophiuchus contains several interesting objects, including Barnard's Star, which is thought possibly to have a planetary system.

Northeast　　　　　**East-Northeast**　　　　　**East**

Vega	0.00 magnitude	+0.5 absolute
VEE-ga	blue-white A0	magnitude
Alpha, α	main sequence V	18h 37m +38° 47′
Lyrae, page 184	26 light-years	August 16 transit
	+0.200 +0.285 −14	STR 982

Location•Brilliant Vega, the third-brightest star seen from midnorthern latitudes, is located 15 degrees to the southeast of Eltanin.

Meaning•The name Vega comes from an Arabian description of this star as the Plunging Eagle.

Lore•Vega is said to have been known in Babylon as the Messenger of Light. The Assyrian title, Judge of Heaven, may have been inspired by its brilliance and high passage on the meridian. Chinese and Japanese astronomical lore described Vega, in Lyra, as the Spinning Damsel. She is seen on the western side of the celestial river, the Milky Way, separated from her lover represented by the star Altair. This tale tells that, once a year, on the seventh night of the seventh moon, a living bridge of magpies rises and spans the celestial river so that the lovers may unite.

The arrival of Vega to a prominent position in the eastern evening sky is a sure sign that the warmest part of the year will soon be upon us.

History•Wilhelm von Struve succeeded in calculating the distance to Vega in 1838, soon after the first stellar distance, that of the star 61 Cygni, had been determined by Friedrich Wilhelm Bessel.

Vega was the first star to be photographed through a telescope. On the night of July 16, 1850, William C. Bond, director of the Harvard Observatory, and J. A. Whipple, a professional photographer, made a 100-second daguerreotype exposure of Vega, using the observatory's 15-inch refractor.

A faint, twelfth-magnitude optical companion to Vega was discovered by Edward E. Barnard on May 21, 1897, the first night astronomers used the great 40-inch refracting telescope of the Yerkes Observatory.

Description•Vega is visible at some time during every night of the year to observers at midnorthern latitudes. It is a blue-white main-sequence star with an absolute magnitude of +0.5. It has 52 times the luminosity of the sun. With an apparent magnitude of 0.00, Vega is one of the few zero-magnitude stars. We observe Vega from a distance of 26 light-years. Its spectrum is dominated by hydrogen absorption lines, and the star's surface temperature is 9,900°K.

Main-sequence stars such as Vega are estimated to have masses about 2.4 times greater than that of the sun. The estimated main-sequence life span for these stars is in the order of magnitude of about 200 million years.

Although Vega has long been used as a standard for the comparison of brightness measurements of other stars, there are suggestions that Vega is actually slightly variable in magnitude. Recent research suggests that Vega shows slight—less than 0.05—magnitude changes.

Vega's prominence in astronomical history was enhanced by the discovery of a system of solid particles in orbit around the star. Knowledge of the existence of this system came as a result of an analysis of data produced by the Infrared Astronomical Satellite *(IRAS)*, which was launched in 1983 as a joint project of the Netherlands, the United Kingdom, and the United States.

Infrared emissions in the vicinity of Vega were attributed to numerous particles, probably similar to the meteoroids and zodiacal dust found in our solar system. Some of this material may still be in the process of accreting into planets; however, planet-sized objects themselves would be difficult to detect with present instruments due to their relatively small surface area compared to that presented by clouds of circumstellar dust.

Nearby Features

EPSILON LYRAE: THE "DOUBLE-DOUBLE" STAR (18h 44m + 39° 40′)

Located just 1.5 degrees to the northeast of Vega is the multiple star system Epsilon Lyrae. Binoculars easily show Epsilon as a pair of fifth-magnitude blue-white stars, having an angular separation equal to nearly 10 percent of the diameter of the full moon. A telescope with an objective diameter of at least 6 inches and a power of 150 magnifications is needed to see the double-double set of four stars.

The more northerly of the two stars of Epsilon visible with binoculars is called Epsilon^1, and its companion is known as Epsilon^2. Epsilon^1 is itself composed of two blue-white main-sequence stars of magnitudes 5.06 and 6.02, with an angular separation of just under 3 arcseconds. The period of revolution for this system is about 1,170 years. Epsilon^2 Lyrae, located 209 arcseconds to the south, is likewise composed of a pair of blue-white main-sequence stars. The separation is about 2 arcseconds, with a period of 585 years.

The four stars of the double-double in Lyra are apparently traveling together through space, since they each have nearly the same proper motion. However, no evidence has been observed of the orbital motion of the Epsilon^1 – Epsilon^2 pair. The period of that orbit is estimated to be about one million years, far too long for astronomers to have seen the stars move along a measurable segment of their orbital path.

The stars of the Epsilon Lyrae system are believed to be about 200 light-years from earth. At this distance, the Epsilon^1 – Epsilon^2 pair has a separation equivalent to 150 times the diameter of the solar system. The true separation of the individual components that compose Epsilon^1 – Epsilon^2 Lyrae are equal to about three and two diameters, respectively, of our solar system.

M57: THE RING NEBULA (18h 53m + 33° 01′)

This is the most beautiful planetary nebula visible through a moderate-size telescope. It looks like a tiny smoke ring about 6 degrees from Vega and about halfway between the stars Beta and Gamma Lyrae. A telescope of at least 6 inches in diameter and 50 power is needed to see the smoky doughnut figure of M57, which is about 75 arcseconds in diameter, with an apparent magnitude of 9.3.

Planetary nebulae are immense shells of gas that are ejected from giant stars in the final stages of their evolution. They received their curious name from a superficial resemblance to planets.

The Ring Nebula in Lyra is believed to be about 1,500 light-years from earth, and the diameter of the ring is about 400 times the size of our solar system. Studies of the ring's velocity of expansion suggest that it was ejected about 20,000 years ago from the star that lies at its center. The ring's eruption evidently carried off most of the central star's original mass, and that star is now a feeble fourteenth-magnitude object with characteristics approaching those of a white dwarf.

BETA LYRAE: AN UNUSUAL VARIABLE STAR (18h 50m +33° 22′)

Beta Lyrae is sometimes called a bright-eclipsing or Lyrid-type variable. At maximum brightness it has a magnitude of 3.4, and its minima alternate between magnitudes 3.8 and 4.1. The brightness variations are caused by common eclipses of two stars of unequal brightness in a common orbit. A complete cycle of brightness changes, from one maximum to the next, occurs during a period of 12.9 days. John Goodricke discovered these magnitude variations in 1784. The changes are most easily noticed by comparing Beta to its neighbor Gamma Lyrae, which has a magnitude of 3.2. At maximum light, Beta and Gamma are nearly equal in brightness; however, at its minimum, Beta is only half the brightness of Gamma.

It has been difficult for astronomers to construct a satisfactory explanation for Beta's characteristics. One model describes the Beta Lyrae binary system as a blue-white giant primary in a very close orbit with a somewhat more massive object, with the two connected by a large stream of gas. A gas disk seems to circle the secondary object, and a vast shell of gas encompasses the entire Beta Lyrae system.

The masses of the components are usually estimated to be about 10 and 20 times that of the sun. The massive secondary object seems to be fainter than its mass should indicate. It has even been suggested that the companion is a black hole, surrounded by a luminous accretion disk. Distance to Beta Lyrae is believed to be about 800 light-years.

Lyra, *the Lyre*

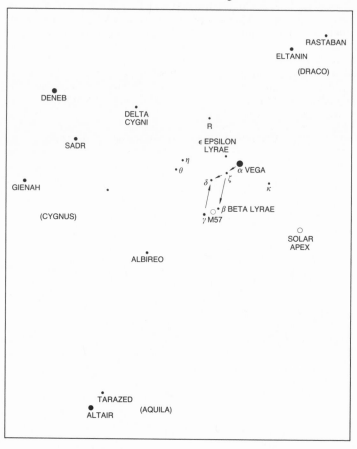

BRIGHT STAR OF LYRA

May 28	Vega	0.0	blue-white	main sequence

LIE-rah
Lyr
Lyrae

Season•June – November

Sky Track•Lyra moves across the sky from northeast to northwest, crossing the meridian overhead.

Lore•Lyra represents the lyre of classical mythology, which was invented by Hermes as a present for his half-brother Apollo. Apollo then gave this instrument to his son Orpheus, musician of the Argonauts, who used it to charm Pluto and his minions in the attempt to free Eurydice from Hades.

Description•The constellation is small, yet it contains some remarkable objects.

THE SOLAR APEX.

The sun and the rest of the solar system is moving through space at a speed of about 20 kilometers per second in the general direction of the star Vega. The imaginary target of this motion is known as the solar apex. The location of this point near Vega was determined through studies of the motions of nearby stars relative to the sun. At its present velocity, the solar system would require about 390,000 years to reach the place now occupied by Vega.

Rasalhague

May 30

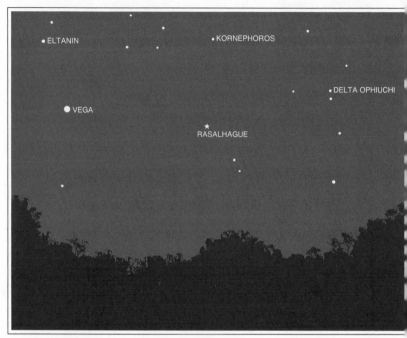

East-Northeast	East	East-Southeast

Rasalhague RAS-al-HAY-gwee Alpha, α Ophiuchi, page 178	2.1 magnitude blue-white A5 giant III 60 light-years +0.117 −0.227 +13	+0.8 absolute magnitude 17h 35m +12° 34′ July 31 transit STR 987

Location•Rasalhague is 17 degrees southeast of Kornephoros and 27 degrees east of Unuk.

Meaning•The name came from a description of the Head of the Serpent Holder.

Lore•Rasalhague marks the head of Ophiuchus. An early Arabian title for this star was the Shepherd. The star Rasalgeti, 5 degrees to the northwest in Hercules, represented the Shepherd's Dog.

An ancient Chinese title for Rasalhague is said to have been the Duke.

Description•Rasalhague is a blue-white giant at a distance of 60 light-years from earth. It has an apparent magnitude of 2.1 and a surface temperature of 8,500°K. This star is about 40 times more luminous than the sun.

The proper-motion path of Rasalhague across the sky displays a wavelike pattern having a period of 8.5 years. This behavior indicates the presence of an unresolved companion star. Analysis of the variations in the path of proper motion suggests masses of about 2.4 and 0.6 that of the sun for the component stars.

Nearby Feature

RASALGETI (17h 15m + 14° 23′)

Rasalgeti, a third-magnitude star, is an interesting double as well as a prime example of a semiregular variable.

Rasalgeti has an average magnitude of 3.5, and the magnitude of its companion is 5.4, with a separation of 4.9 arcseconds between the two stars. Their colors are yellow-orange and yellow-white. The secondary is a spectroscopic binary. Magnifications above 150× may be needed to resolve the components of Rasalgeti.

Rasalgeti, the primary component, is an example of a red giant and has a semiregular period of brightness variation of about six years, and a range between 3.0 and 4.0 magnitude. There are also smaller variations of brightness with periods of 50 to 130 days.

The star is surrounded by an envelope of gas whose radius is about 25 times the size of the orbit of the planet Pluto in our solar system. Rasalgeti is evidently pumping material into this envelope at a substantial rate. Such material, produced by fusion reactions in red giants, may account for much of the enrichment of the Milky Way's disk by heavy elements.

Studies of the binary system indicate a mass 14 times that of the sun for Rasalgeti. Measurements with a device known as a stellar interferometer reveal its average diameter to be about 500 times that of the sun. The apparent diameter of this star is 0.030 arcsecond, which is equal to the parallactic shift in a star's position when seen at a distance of 110 light-years. This minute angle is close to the current limit for accurate measurement of distances on the celestial sphere.

See the constellation maps for Hercules (page 158) and Ophiuchus (page 178).

June Stars

Cheleb

June 7

East	East-Southeast	Southeast

Cheleb	2.8 magnitude	+0.36 absolute
CHAY-leb	yellow K2	magnitude
Beta, β	giant III	17h 43m +04° 34′
Ophiuchi, page 178	100 light-years	August 2 transit
	−0.042 +0.159 −12	STR 1020

Location•Cheleb is found 8 degrees south of Rasalhague in the constellation Ophiuchus. It is a third-magnitude star.

Meaning•The name comes from an Arabic description of the Shepherd's Dog.

Description•At a distance of 100 light-years from earth, the yellow giant star Cheleb shines in our night sky with an apparent magnitude of 2.8. Its luminosity is 60 times greater than that of the sun. Cheleb's surface temperature is estimated to be 4,200°K.

Nearby Feature

BARNARD'S STAR (17h 58m +04° 40′)

Located 3.5 degrees to the east of Cheleb, this star has the highest observed rate of proper motion, 10.3 arcseconds per year. Named after its discoverer, Edward E. Barnard, Barnard's Star is an orange dwarf with an absolute magnitude of +13.4 and is nearly 3,000 times less luminous than the sun. As a result, it is seen in the sky with an apparent magnitude of only 9.5, despite its relatively close distance of 5.9 light-years. The only known stars closer to earth are the sun and members of the Alpha Centauri system.

In four years' time, Barnard's Star traverses a portion of the celestial sphere approximately equal to the apparent diameter of the planet Jupiter. In 180 years, it moves a distance equivalent to the angular diameter of the full moon.

Barnard discovered this remarkable star while comparing photographs taken of this region of sky in the years 1894 and 1916. The star's extraordinary rate of proper motion results from a high intrinsic velocity combined with its proximity to earth. Only three other stars have proper motions of more than 5 arcseconds per year. Most stars are too distant for proper-motion measurements, and of those that have been determined, the vast majority are in the range of less than 1 arcsecond per year. It is no wonder that the rapid dwarf in Ophiuchus is sometimes called Barnard's Runaway Star.

Astronomer Peter van de Kamp of the Sproul Observatory at Swarthmore College in Pennsylvania has studied Barnard's Star's proper-motion path over a period of many decades. He has suggested that oscillations in the star's path may indicate the presence of two unseen planetary companions.

See the constellation map for Ophiuchus, page 178.

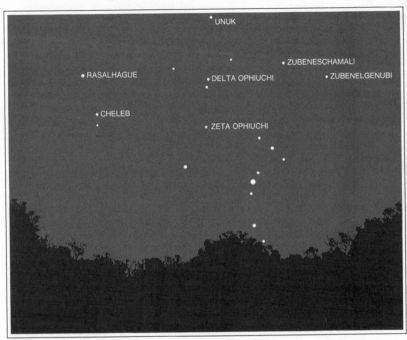

East-Southeast   **Southeast**   **South-Southeast**

Zeta Ophiuchi ZAY-tah oh-fee-YOU-kye Zeta, ζ Ophiuchi, page 178	2.6 magnitude blue-white O9.5 main sequence V 600 light-years +0.012 +0.023 −15	−3.6 absolute magnitude 16h 37m −10° 34′ July 17 transit STR 1024

Location•Zeta Ophiuchi is a third-magnitude star found 9 degrees southeast of Delta Ophiuchi.

Lore•Zeta, at times, shared the name Sabik with Eta Ophiuchi. The name meant "the preceding ones," who rose before the Scorpion.

In China the star was known as Han, the name of an ancient dynasty.

A Persian description for Zeta Ophiuchi was the Bound One, indicating it was wrapped in the coils of the Serpent.

Description•Zeta is too far away for accurate distance measurement by the method of trigonometric parallax. The distance estimate of 600 light-years is based on the comparison of Zeta's apparent and absolute magnitudes. Apparent magnitudes are measured through the techniques of photometry, in which photoelectric devices help to determine a star's brightness. The absolute magnitude is estimated from a comparison with other stars having similar spectral characteristics. Zeta Ophiuchi is rather close to the plane of the Milky Way and also lies at a considerable distance from the earth. As a result, absorption of its light by interstellar dust must be taken into account when calculating its distance. This dust causes us to see Zeta about 1 magnitude fainter than if the star were located in a region of the sky 90 degrees from the Milky Way's plane.

The method used for estimating Zeta Ophiuchi's distance is known as spectroscopic parallax. This technique is a vital link in the set of procedures used for estimating the distances of objects in the universe.

Zeta Ophiuchi shines with an apparent magnitude of 2.6, and it has a luminosity nearly 3,000 times that of our sun. The surface temperature is believed to be 32,000°K, which is typical for a blue-white main-sequence star of this type. The estimated mass is about 20 times that of the sun. Such a star would have a main-sequence life expectancy in the order of 100 million years.

Nearby Feature

SCORPIUS X-1 (16h 17m $-15°$ 31′)

Seven degrees to the southwest of the star Zeta Ophiuchi is located the strongest concentrated source of X rays yet discovered in the sky. It was also the first X-ray source found in the constellation Scorpius and therefore was called Scorpius X-1. This object was identified in 1962 with rocket-borne instruments. By 1966, Scorpius X-1 was thought to be somehow related to a hot, blue-white star of the thirteenth magnitude seen in the same location. This star's distance is estimated to be 1,600 light-years, and it has characteristics somewhat similar to those of an old nova.

The radiation from Scorpius X-1 carries about 1,000 times more energy than the visible light from the object. Several models have been suggested to account for this strong and very concentrated X-ray source. The most widely held theory describes the object as a dense, collapsed member of a binary-star system. This shrunken star is said to be surrounded by a shell or accretion disk formed from material drawn away from its companion star. As surrounding gas falls towards the collapsed stellar remnant, the gas accelerates, and in the process it is heated to temperatures sufficient for the generation of the powerful X rays that are observed from earth.

See constellation maps for Ophiuchus (page 178) and Scorpius (page 198).

Delta Cygni   June 11

North-Northeast	Northeast	East-Northeast

Delta Cygni	2.9 magnitude	+0.3 absolute
DEL-tah sig-NYE	blue-white B9.5	magnitude
Delta, δ	subgiant IV	19h 45m +45° 08′
Cygni, page 194	110 light-years	September 2 transit
	+0.049 +0.049 −20	STR 1035

Location•As we approach the longest days of the year, the stars of the constellation Cygnus begin to move into position on the Sky Screen as evening twilight fades. Delta Cygni, the first star featured in this constellation, is found 13 degrees to the northeast of Vega. It is a third-magnitude object that marks the western end of the traverse of the Northern Cross.

History•In 1830, Wilhelm von Struve discovered Delta Cygni's sixth-magnitude companion star.

Description•The Delta Cygni binary system includes a blue-white primary along with a somewhat cooler blue-white main-sequence star. The primary is classified as a subgiant. The separation of the pair is about 2 arcseconds, and the period of revolution is estimated at 540 years. Delta Cygni and its companion are about 110 light-years from earth.

The primary has a surface temperature of 10,500°K and 63 times the sun's luminosity.

Nearby Features

VARIABLE STAR RR LYRAE (19h 25m +42° 50′)

RR Lyrae is a white star that ranges between magnitude 7.1 and 8.0 over a period of 0.57 day. It is 4 degrees to the southwest of Delta Cygni. Because this star is an eighteenth-magnitude object, finding it requires a telescope and a rather detailed map of this part of the sky.

RR Lyrae is the best prototype of an important class of variable stars called RR Lyrae stars or cluster variables. These stars are typically found in the Galactic halo and in globular clusters and are therefore a type of population II star. RR Lyrae itself happens to be passing through the plane of the Milky Way, but as a population II object, it has a much different space motion from the disk stars such as the sun, which are orbiting the Milky Way's center, always keeping close to the plane of the Galaxy's disk. RR Lyrae's distance is estimated to be about 900 light-years from earth.

RR Lyrae stars are pulsating variables, as are the Delta Cephei variables. However, RR Lyrae stars have lower luminosities and shorter periods of light variation than do the Cepheids. In most cases, RR Lyrae stars vary in absolute magnitude between 0.0 and +1.0 during periods of less than one day.

See the following constellation map.

CYGNUS A RADIO SOURCE (19h 58m +40° 36′)

Delta Cygni has another interesting neighbor, known as Cygnus A. This object was discovered in 1948, when radio astronomy was just beginning to flower. It is one of the most powerful sources of radio energy in the heavens. During the year 1951, Cygnus A was associated with a pair of eighteenth-magnitude images photographed by the 200-inch, 5-meter Hale telescope on Mount Palomar. At first, the radio energy was believed to be produced as a result of the collision of two remote galaxies.

At the present time, the Cygnus A radio energy is attributed to an unusual giant elliptical galaxy, estimated to be between 500 million and one billion light-years from earth.

Cygnus A is located 4.5 degrees to the southeast of the star Delta Cygni. See the following constellation map.

Cygnus, *the Swan*

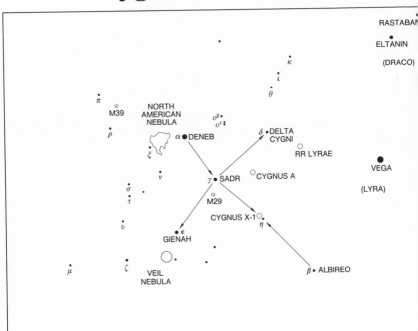

BRIGHT STARS OF CYGNUS

June 11	Delta Cygni	2.9	blue-white	subgiant
June 23	Sadr	2.2	white	supergiant
June 25	Deneb	1.3	blue-white	supergiant
July 3	Gienah	2.5	yellow	giant

SIG-nus
Cyg
Cygni

Evening Season•June – January

Sky Track•Cygnus rises in the northeast, crosses the meridian overhead, and sets towards the northwest.

Lore•The stars of Cygnus outline an attractive asterism often known as the Northern Cross.

The characterization of these stars as a bird began in the ancient Middle East and may have been related to the legend of the roc, the great bird encountered by Sinbad the Sailor in the *Arabian Nights*.

In Greece and Rome, the constellation was popularly known as the Bird. The present name was applied in commemoration of several young men in classical mythology, all named Cycnus *[sic]*, who were changed into swans.

In Arabia, the constellation was usually called the Hen. The traverse of the Northern Cross was called the Riders.

Description•Deneb marks the Swan's tail and Albireo marks its head. The bird is portrayed flying towards the southwest, directly along the centerline of the Milky Way.

Nearby Feature

CYGNUS X-1: A POSSIBLE BLACK HOLE (19h 56m + 35° 04′)

Chinese records indicate that a bright "guest star" was observed in the constellation Cygnus, between Sadr and Albireo, on the night of October 24, 1408. This object may have been the supernova that produced the powerful X-ray source known as Cygnus X-1, located about one degree to the northeast of fourth-magnitude Eta Cygni.

Cygnus X-1 has, for some time, been suspected of being a black hole. It was first detected by X-ray sensors sent aloft by rocket in 1962. In 1971, the X-ray satellite *Uhuru* found great variations in the intensity of this source over periods as brief as a few hundredths of a second. This indicated to astronomers that the object was remarkably small.

A supergiant star of the ninth magnitude was later identified as a binary companion of the Cygnus X-1 source. Additional research has shown that this supergiant is located about 7,000 light-years from earth and that its dark companion has a mass between 10 and 20 times that of the sun. The pair has an orbital period of 5.6 days. There is also evidence of quantities of hot gas being transferred from the supergiant to the unseen companion.

This combination of great mass with a very small diameter suggests to astronomers that Cygnus X-1 represents a supernova remnant that has collapsed to form a black hole. Such an object would have a concentration of matter and gravity that prevents the escape of light from its vicinity.

X rays that suggest the presence of this object are probably produced as gas is pulled away from the neighboring supergiant and then accelerated as it approaches the black hole. When this gas increases sufficiently in velocity and temperature, X rays are produced.

The black hole itself is surrounded by a spherical region known as the event horizon, which marks the limits of the object's visibility. From beneath its surface, nothing, including light, can escape due to the intense gravitational field. In 1916, the physicist Karl Schwarzschild found that the radius of such an event horizon around a small massive object would be approximately equal to three kilometers for every solar mass of material that the object contains.

Graffias   June 18

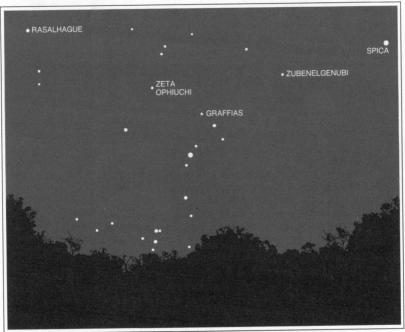

| Southeast | South-Southeast | South |

Graffias GRAF-ee-as Beta, β Scorpii, page 198	2.5/2.6 magnitude blue-white B0.5 main sequence V 600 light-years −0.006 −0.021 −1	−3.7 absolute magnitude 16h 05m −19° 48′ July 8 transit STR 1066

Location·Graffias is the first of the stars featured in Scorpius to appear in the southeastern sky. It is a third-magnitude object located 12 degrees southwest of Zeta Ophiuchi and 11 degrees north of the first-magnitude star Antares.

Meaning·The name is of Greek origin and evidently referred to a Crab, this creature being substituted for the Scorpion.

Lore·In Arabia, Graffias marked the Crown of the Forehead of the Scorpion. Along with nearby Delta and Pi, this star was known as the Row or Ridge in the star lore of India. It has been said that in China, Graffias was called the Four-horse Chariot of Heaven and was revered by horsemen.

Description·Graffias is a fine double star for observations with a small telescope. The components have magnitudes of 2.6 and 4.9 with a separation of 13.7 arcseconds. Both components are blue-white main-sequence stars. A sixth-magnitude companion was discovered by S. W. Burnham at a distance of 0.5 arcsecond.

Speckle interferometry and occultations by the planet Jupiter, as well as by the moon, indicate the presence of additional stars in the Graffias system.

An occultation is a form of eclipse, in which the light from a celestial object is blocked from our view by a solar-system body such as a planet, satellite, or asteroid. Studies of stars that are being occulted may reveal very close, unknown companions. The light from a single star would be expected to vanish abruptly when eclipsed by the airless edge of the moon. Light from the secondary star in a close binary system may cause the image to persist for a fraction of a second beyond what would be expected for a single star.

Speckle interferometry is a technique devised during the 1970s that is used to study close double stars and to partly resolve the disks of some supergiant stars.

The stars in the Graffias system have an estimated distance of about 600 light-years. Graffias, the primary, is a massive blue-white main-sequence star. It is estimated to be nearly 20 times the size of the sun. Such a star races through its main-sequence life span over 100 times faster than does the sun.

Graffias is about 2,800 times the luminosity of the sun and has a surface temperature of about 27,000°K. The star is one of the bright members of the Scorpius-Centaurus star association. Members of this group formed at approximately the same time from a common molecular cloud system.

Nearby Feature

NU SCORPII: A QUADRUPLE STAR (16h 12m − 19° 27′)

The quadruple star Nu Scorpii is located just 1.5 degrees to the east of Graffias. A small telescope reveals a pair of blue-white stars with magnitudes of 4.0 and 6.3 with a separation of 41 arcseconds. Telescopes with diameters of 10 inches or more may further resolve the pair into a total of four components. The two secondary pairs have individual separations of about 1.0 and 2.0 arcseconds respectively.

Scorpius, *the Scorpion*

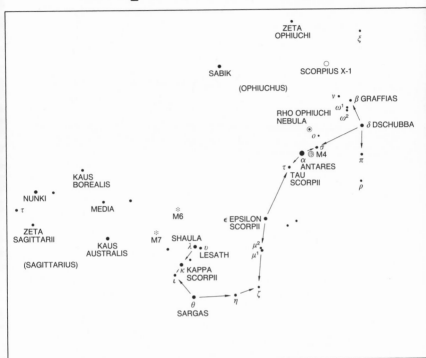

BRIGHT STARS OF SCORPIUS

June 18	Graffias	2.5	blue-white	main sequence
July 7	Dschubba	2.3	blue-white	main sequence
July 15	Antares	1.0	yellow-orange	supergiant
July 20	Epsilon Scorpii	2.3	yellow	giant-subgiant
July 30	Lesath	2.7	blue-white	subgiant
July 31	Shaula	1.6	blue-white	subgiant
August 1	Sargas	1.9	blue-white	supergiant
August 2	Kappa Scorpii	2.4	blue-white	subgiant

SKOR-pea-uss
Sco
Scorpii

Evening Season•June – August

Sky Track•Scorpius contains several of the most southerly of the bright stars seen from the continental United States. Because of the location of these stars, Scorpius is seen near to the horizon as it moves from southeast to southwest.

Lore•The curving line of bright stars in Scorpius clearly resembles a scorpion. The ancient Greeks usually included the stars of Libra, which formed the Scorpion's claws, in the constellation. This extended zodiacal figure was sometimes called the Slayer of Orion, since in one version of the myth Orion was stung to death by a scorpion. Hostility between these two figures is suggested by the fact that Orion sets in the west just as the stars of Scorpius rise in the southeast.

The usual association of these stars with a Scorpion rather than a more benign, similarly shaped creature such as a crab or lobster has been attributed to Mesopotamian times, when the apparent weakening of the sun during autumn occurred as it moved into this part of the sky.

Early Chinese astronomers included these stars as a part of a large figure known as the Azure Dragon.

Description•This large and bright zodiacal constellation is a visual treat; however, you need a nearly clear view of the southern horizon in order to appreciate its beauty fully.

Many of the bright stars of Scorpius are actually related to each other, having been formed in a vast molecular cloud complex less than 20 million years ago. They belong to a group of over 100 stars that form the great Scorpius-Centaurus association, which was first described by Jacobus C. Kapteyn of the Netherlands in 1914. These stars are distributed across a wide portion of sky centered on the constellations Scorpius and Centaurus and, at an average distance of about 500 light-years, are the closest example of this type of star group. They are members of an OB association of stars and as such are young, massive, and highly luminous.

Antares is the brightest star in the Scorpius-Centaurus association, and it is probably the most massive since it alone has evolved to the red giant stage. Most other members of the association are blue-white stars of spectral type B.

Alderamin   June 21

North   **North-Northeast**   **Northeast**

Alderamin	2.4 magnitude	1.5 absolute
al-DARE-ah-min	blue-white A7	magnitude
Alpha, α	subgiant – main	21h 19m +62° 35′
Cephei, page 202	sequence IV-V	September 26
	50 light-years	transit
	+0.150 +0.052 −10	STR 1076

Location•Alderamin is the brightest star in the rather faint constellation of Cepheus and is located 18 degrees from Deneb. Alderamin is a second-magnitude star.

On this day, the sun is nearing the most northern point of the ecliptic, near the third-magnitude star Mu Geminorum. This location on the ecliptic is known as the summer solstice, and when the sun reaches it, summer begins in the northern hemisphere.

Meaning•The origin of the name Alderamin is obscure.

Lore•The Tartar astronomer Ulug-Beg called Alderamin and several of its neighbors the Stars of the Flock.

Description•Alderamin is a blue-white star that is variously described as either a subgiant or a main-sequence object. It is 50 lightyears from earth. The luminosity of Alderamin is 21 times that of the sun, and its surface temperature is estimated to be about 8,000°K.

Nearby Feature

DELTA CEPHEI: VARIABLE AND DOUBLE STAR (22h 29m + 58° 25')

Variable star Delta Cephei is a fourth-magnitude object of class F5Ib-G2Ib found 9 degrees from Alderamin. This pulsating supergiant was the first of its type to be discovered. Such stars are called Cepheid variables after their prototype, and they serve astronomers as a vital link in the set of methods used to determine distances in the universe.

The brightness of Delta Cephei ranges between magnitudes 3.4 and 4.3 over a period of 5 days, 8 hours, and 48 minutes. The magnitude range and period are both slightly variable.

Delta Cephei is also an attractive double star, which may be observed with a small telescope. The components are white to yellow-white and blue-white. The average magnitude of Delta is 3.9 and that of its companion is 6.2. The pair has a separation of 41 arcseconds.

See the following constellation map.

In 1912, Henrietta Leavitt, an astronomer at Harvard College Observatory, discovered a relationship between the apparent magnitudes of Cepheid variables in the Magellanic Clouds and their periods of light variation. The Cepheids with the greater average brightnesses were seen to have longer periods between their maximum and minimum brightness than did less bright Cepheids. The Magellanic Clouds are small satellite galaxies associated with the Milky Way. Since their size is small in comparison with their distance from earth, all stars in these galaxies are roughly the same distance from earth. Hence, differences in brightness of Magellanic Cloud stars, as viewed from earth, result mainly from actual luminosity differences between these stars. Leavitt was therefore able to correlate periods of variability with the relative magnitudes of many Cepheid variables that she found in the Magellanic Clouds. This correlation is known as the period-luminosity relation.

During 1914, Harlow Shapley of the Mount Wilson Observatory used a method known as statistical parallaxes to measure the distances and estimate absolute magnitudes of several Cepheids relatively near to the earth. In this technique, a star's proper motion and radial velocity are measured and used as clues to its distance. As a result of his study, Shapley was able to calibrate the Cepheid period-luminosity relation. He soon began to use Cepheid variables to help measure distances around the Milky Way galaxy.

Cepheids are pulsating giant stars with periods of about 3 and 50 days and a range of average absolute magnitudes from −1.5 to −5.0.

Cepheus, *the King*

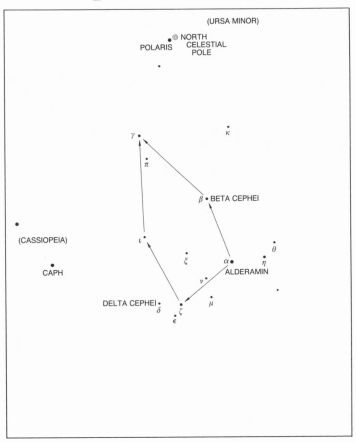

(URSA MINOR)

POLARIS ⊚ NORTH
CELESTIAL
POLE

κ

γ

π

β • BETA CEPHEI

(CASSIOPEIA)

ι

ζ

α • ALDERAMIN

θ

η

CAPH

ν

δ DELTA CEPHEI

ζ

μ

ε

BRIGHT STAR OF CEPHEUS

| June 21 | Alderamin | 2.4 | blue-white | subgiant – main sequence |

SEE-fee-us
Cep
Cephei

Cepheus is a faint constellation situated between Draco and Cassiopeia in the northern sky. Its shape resembles the end of a house with a pointed roof, the top of which is towards the north.

Evening Season•As a circumpolar constellation, Cepheus is above the horizon at all times of night throughout the year. During the months June through February, Cepheus is away from the horizon and most easily seen during the evening hours.

Sky Track•Northern circumpolar

Lore•In classical myths, Cepheus is said to portray a king of Ethiopia. He was traditionally shown with arms outstretched, one foot on the North Celestial Pole and the other along the solstitial colure.

In Greek mythology, Cepheus was the husband of Queen Cassiopeia and the father of Andromeda. Together they are known as the Royal Family of the Sky, and their story also includes the adventures of Perseus and the flying horse Pegasus, as well as the sea creature Cetus. These constellations occupy a significant part of the evening sky during the autumn season.

Early Arabians saw some of the stars of Cepheus as Sheep that were associated with a large celestial Pen, which, in their mythology, filled a large part of the northern sky.

Nearby Features

BETA CEPHEI: VARIABLE STAR (21h 29m +70° 34′)

Beta Cephei is an important variable star found in the constellation Cepheus in the vicinity of Alderamin. Beta is located 8 degrees north of Alderamin and, as is the case with Delta Cephei, it is a prototype for a class of variables as well as being a small-telescope double.

The period of Beta Cephei is about 4 hours and 34 minutes, and its brightness varies between magnitudes 3.16 and 3.27. It is a pulsating blue-white subgiant star at a distance of about 1,000 light-years from earth.

Beta Cephei – type variables have only slight changes in brightness and possess periods of less than one day. They were first identified as a result of Doppler shifts in their spectral lines caused by the pulsating motion of their surfaces as these stars expand and contract.

Beta Cephei stars are spectral-type-B blue-white subgiants that have just begun to evolve away from the main sequence. Mirzam in Canis Major and Shaula in the constellation Scorpius are bright examples of this type of variable, of which several dozen are known.

There is an eighth-magnitude blue-white companion to Beta Cephei that lies at a distance of 14 arcseconds from the primary. The pair offers a very appealing contrast in brightness and is separated at about 100 magnifications.

HERSCHEL'S GARNET STAR (12h 44m +58° 47′)

Mu Cephei is a fourth-magnitude star located 5 degrees to the southeast of Alderamin. It is a supergiant semiregular variable of spectral type M2. Despite the description provided by William Herschel, the spectral class indicates a yellow-orange apparent color, and this hue may be observed with the help of a telescope.

Sadr

Northeast   **East-Northeast**   **East**

Sadr	2.2 magnitude	−4.7 absolute
sad-der	white F8	magnitude
Gamma, γ	supergiant Ib	20h 22m +40° 15′
Cygni, page 194	800 light-years	September 12
	+0.001 +0.002 −8	transit
		STR 1085

Location•Sadr marks the center of the Northern Cross, an attractive asterism formed from the brightest stars in the constellation Cygnus. The star Sadr is found 6 degrees south of Deneb and 20 degrees east from Vega.

Meaning•The name Sadr is of Arabic origin, from a description of the Breast of the Hen.

Description•Sadr is a white supergiant, with a distance from earth estimated to be 800 light-years. It is 6,300 times more luminous than the sun and has a surface temperature of 6,000°K.

Nearby Features

ALBIREO: A DOUBLE STAR (19h 31m +27° 58′)

Albireo is one of the loveliest and most frequently observed double stars. The primary Albireo A is itself a double and displays spectral lines characteristic of a yellow giant and a blue-white main-sequence star. These two stars are in a very close orbit and the light of the giant dominates, making Albireo A appear yellow, with an apparent magnitude of 3.1. The visual secondary, Albireo B, is a blue-white main-sequence star having a magnitude of 5.1. The separation between components A and B is 34.7 arcseconds.

The components of Albireo have a very similar space motion and are believed to be a binary pair, although no orbital motion has been observed since the measurements of Wilhelm von Struve in 1832.

Try placing the eyepiece of your telescope slightly out of focus in order to observe more easily the color contrast between Albireo A and Albireo B. This technique presents the stars as small colored disks rather than points of light.

See the Cygnus constellation map, page 194.

THE NORTH AMERICAN NEBULA (20h 57m +44° 08′)

A pair of binoculars is sufficient to give you a glimpse of the North American Nebula if the night is clear and you are far from artificial lights. This cloud of luminous gas and obscuring dust is about 1,600 light-years from earth, and its shape is reminiscent of the outline of the North American continent.

The nebula grows with visible light as a result of the stimulation of its hydrogen gas by ultraviolet radiation from Deneb and other nearby stars. The North American Nebula covers an area about 1.5 degrees in diameter, and it is centered 3 degrees to the east of Deneb.

This nebula was discovered and named by Max Wolf of the Heidelberg Observatory in 1891. It was found through the use of wide-angle, long-exposure photography.

See the Cygnus constellation map, page 194.

THE MILKY WAY IN CYGNUS

There are marvelous fields of Milky Way stars and nebulae in the regions around Cygnus. These fields of distant stars are best seen through binoculars or a low-power telescope.

Situated between Sadr and Albireo is the great Cygnus Star Cloud, composed of countless stars in one of the Milky Way's spiral arms, as seen from a distance of about 7,000 light-years. The nineteenth-century astronomer Thomas W. Webb, in describing this region, noted that "its fields in low power are overpowering in magnificence." The Great Rift of the Milky Way begins in Cygnus and is seen to extend towards the southwest.

Deneb

North-Northeast	Northeast	East-Northeast

Deneb	1.3 magnitude	−7.3 absolute
DEN-ebb	blue-white A2	magnitude
Alpha, α	supergiant Ia	20h 41m +45° 17m
Cygni, page 194	1,600 light-years	August 2 transit
	+0.001 +0.005 −5	STR 1091

Location•Deneb is a first-magnitude star located at the top of the Northern Cross, about 22 degrees to the east of Vega.

Meaning•The name Deneb comes from an Arabic description meaning "the tail of the hen."

Lore•Deneb marks the northeast corner of the asterism known as the Summer Triangle. Vega in the constellation Lyra, and Altair in Aquila, are the other stars in this prominent feature of summer evening skies.

In the constellation figure of Cygnus, Deneb represents the swan's tail.

Description•Deneb is one of the most luminous stars seen in the sky. Because of its high intrinsic brightness, it appears to us as a first-magnitude object, even at the substantial distance of 1,600 light-years. From 4.4 light-years, the distance of Alpha Centauri, nearest known star system beyond the sun, Deneb would appear on earth as bright as the full moon.

Deneb is a blue-white supergiant with an absolute magnitude of −7.3 and a luminosity 70,000 times greater than that of the sun. Its surface temperature is about 9,000°K. In order for a star to shine as brilliantly as Deneb, it is estimated that a mass about 25 times greater than that of the sun is required.

Nearby Feature

THE CYGNUS SUPERBUBBLE (CENTER: 20h 40m +40° 00′)

During the late 1970s, the first High Energy Astronomy Observatory, *HEAO-1,* made a comprehensive survey of X-ray emissions from space. While this observatory was in orbit around the earth, it provided data that enabled astronomers to discover a remarkable object, the Cygnus Superbubble.

The *HEAO-1* satellite revealed a vast ring of X-ray emissions, with an apparent diameter of about 15 degrees, in the sky around the stars Deneb and Sadr in the constellation Cygnus. A gap in the ring's southwestern side is caused by the absorption of X rays by the Great Rift of the Milky Way, which is located between earth and the ring.

This X-ray ring is believed to represent our cross-sectional view of the extremely hot surface of an enormous bubble that has been blown within an interstellar molecular cloud. The X-ray spectrum suggests a temperature of about 2 million degrees Kelvin for the surface of the Cygnus Superbubble. The Superbubble is at an estimated distance of 6,000 light-years from earth and it has a diameter of 1,300 light-years. It is one of the largest structures known in the Milky Way Galaxy.

Near the center of the bubble is a group of extremely luminous young stars known as the Cygnus OB2 association. Ultraviolet radiation from these stars flows into space, heats the surrounding gas, and stimulates it to send out radio emissions over a region 500 light-years in diameter. This region surrounding the OB2 association has been known since the 1950s and is designated as the Cygnus X radio source.

Sabik

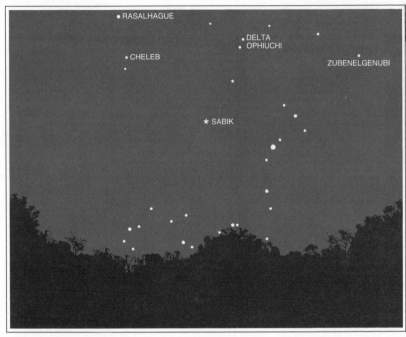

East-Southeast	Southeast	South-Southeast

Sabik	2.4/3.0 magnitude	1.4 absolute
SAY-bik	blue-white A2	magnitude
Eta, η	main sequence V	17h 10m −15° 43′
Ophiuchi, page 178	70 light-years	July 25 transit
	+0.038 +0.095 −1	STR 1091

Location•Second-magnitude Sabik is located 10 degrees southeast of Zeta Ophiuchi and 15 degrees east of Antares. It is in the southern portion of the constellation Ophiuchus.

Meaning•The name Sabik is derived from a description indicating that this star preceded the rise of the constellation Scorpius.

Lore•Sabik traditionally was portrayed as marking the right knee of Ophiuchus, and it marks the junction of that constellation with Serpens, the Serpent.

History•Sabik is a close visual binary and was first resolved by S. W. Burnham using the 36-inch refracting telescope at Lick Observatory in 1889, the year after this instrument was placed in operation.

Description•Sabik is composed of two nearly identical stars having a combined apparent magnitude of about 2.4 and a separation of less than 1 arcsecond.

Both components are blue-white main-sequence stars with individual magnitudes of 3.0 and 3.5. The pair's distance from earth is 70 light-years.

The components of Sabik have 23 and 17 times the sun's luminosity and slightly more than twice the sun's mass. Such stars are estimated to have a main-sequence life expectancy in the order of magnitude of about 500 million years.

July Stars

Gienah

Northeast **East-Northeast** **East**

Gienah	2.5 magnitude	+0.6 absolute
GEE-nah	yellow K0	magnitude
Epsilon, ϵ	giant III	20h 46m +33° 58′
Cygni, page 194	75 light-years	September 18
	+0.355 +0.329 −11	transit
		STR 1123

Location•Gienah marks the eastern end of the traverse of the Northern Cross. It is a second-magnitude star located about 10 degrees to the southeast of Deneb.

Meaning•The name originated with an Arabic word for "wing" and, without the *h*, is also applied to the star Gamma Corvi.

Description·Gienah is a yellow giant star at a distance of about 75 light-years from earth. It is estimated to be 50 times more luminous than the sun and to have a surface temperature of about 4,500°K.

There is a twelfth-magnitude optical companion at a distance of 55 arcseconds. A thirteenth-magnitude companion, which has the same proper motion as Gienah and is therefore probably gravitationally bound to the primary, lies at a distance of 78 arcseconds.

Nearby Features

61 CYGNI: AN HISTORIC AND ATTRACTIVE DOUBLE STAR
(21h 07m +38° 45′)

Located 6 degrees to the northeast of Gienah, this double can be resolved with a small telescope. Its white and yellow components have magnitudes of 5.2 and 6.0, with a separation of 29 arcseconds.

In 1838, the astronomer and mathematician Friedrich W. Bessel, using a special 6¼-inch telescope at the Königsberg Observatory in East Prussia, was able to measure the parallax and thereby calculate the distance of 61 Cygni. This was the first time that the distance of a star other than the sun had been determined.

61 Cygni was selected as a likely subject for Bessel's study because it has the relatively large proper motion of 5.2 arcseconds per year, which suggested that it is one of the closest stars to earth.

THE VEIL NEBULA: A SUPERNOVA REMNANT (20h 54m +31° 30′)

Photography reveals a beautiful, circular supernova remnant known as the Veil Nebula, which is centered 3 degrees to the southeast of Gienah.

This graceful object was discovered by William Herschel in 1784, with the help of his 18-inch reflecting telescope. Although with fine observing conditions the nebula may be glimpsed through a small telescope using low power, photographs are required in order to see the delicate curving filaments that characterize the Veil Nebula.

The name Cygnus Loop is sometimes used for the nebula, and the names Filamentary Nebula and Network Nebula are often applied to its major parts. The Veil Nebula has a diameter of about 2.5 degrees, which is 25 times the area of the full moon's disk.

Measurements of the expansion rates of the nebula's filaments indicate that it originated about 65,000 years ago, when a massive star exploded as a supernova.

Beautiful filaments, which were formed as gas from the outer layers of the exploding star expanded and pushed into the surrounding interstellar medium, are observed as the Cygnus Loop.

See the Cygnus constellation map, page 194.

East-Northeast   **East**   **East-Southeast**

Tarazed	2.7 magnitude	−2.3 absolute
TAR-ah-zed	yellow K3	magnitude
Gamma, γ	giant II	19h 46m +10° 37′
Aquilae, page 214	350 light-years	September 3 transit
	+0.016 +0.002 −2	STR 1124

Location•Tarazed is a third-magnitude star located 25 degrees of arc southwest of Gienah. It is just 2 degrees northwest of the first-magnitude star Altair in the constellation Aquila.

Meaning•The name Tarazed was taken from a Persian description of the star and its two neighbors, Altair and Alshain, as the Scale Beam.

Tarazed is one of many star names devised in the early nineteenth century by the astronomer Giuseppi Piazzi.

Description•Tarazed is a yellow giant star that shines in the sky with an apparent magnitude of 2.7. It is nearly two magnitudes fainter than its neighbor Altair, which appears 6 times brighter than Tarazed.

The distance of Tarazed is about 350 light-years, and this star has a luminosity 700 times greater than the sun's. Tarazed's surface temperature is estimated to be about 4,000°K.

Nearby Features

GAMMA DELPHINI: DOUBLE STAR (20h 47m + 16° 07')

Gamma Delphini is located 15 degrees northeast of Tarazed in the faint constellation Delphinus, the Dolphin. The components of Gamma have magnitudes of 4.3 and 5.1 with a separation of 10 arcseconds. The colors are yellow and white, and the pair is a fine sight in a small telescope. Gamma Delphini's components are about 125 light-years from earth.

See the following constellation map.

SS 433 (19h 09m +4° 54')

Several years ago, there were reports that astronomers had discovered an object that showed evidence of high-speed motion in opposite directions at the same time. This strange object is designated SS 433, and it lies near the central plane of the Milky Way at a distance of about 11,000 light-years from earth. Its location in the sky is 11 degrees to the southwest of Tarazed; however, it is too distant to be seen with a small telescope.

One explanation for the very high speed, oppositely directed motion detected in SS 433 relies on a binary-star model. It has been suggested that SS 433 is a binary system that contains both a giant star and a shrunken supernova remnant, probably a neutron star. Gas from the giant's atmosphere is drawn towards the small, dense companion by virtue of the small object's intensely concentrated gravitational field. As the gas is transferred, it begins to orbit rapidly and form an accretion disk around the neutron star. In some way, gas is propelled into space from both the top and bottom of the accretion disk, in opposite directions, at speeds of about 25 percent of the speed of light. The precession of the accretion disk's axis results in both red- and blue-shifted light reaching earth from the two beams. It is these spectral shifts that account for our perception of both high speed and opposite motion from SS 433.

The complete understanding of these beams, as well as the nature of variable radio and X-ray emissions from SS 433, still remains a mystery.

See the following constellation map.

Aquila, *the Eagle*

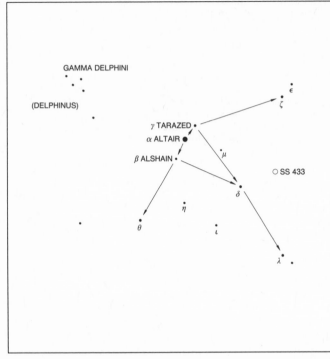

BRIGHT STARS OF AQUILA

July 4	Tarazed	2.7	yellow	giant
July 6	Altair	0.8	blue-white	main sequence

ACK-will-lah
Aql
Aquilae

Evening Season•June – December

Sky Track•This is a constellation that lies along the celestial equator. Its stars rise close to the eastern point on the horizon, pass the meridian about halfway between the horizon and the zenith, and set directly towards the west.

Lore•Traditional star atlases have portrayed Aquila as an eagle flying towards the east, through the Milky Way.

The Romans considered Aquila to be a representation of the Eagle of Jupiter. It was said that this Eagle helped Jupiter in his struggle with the Titans for control of the universe. The bird also carried the Trojan prince Ganymede to Mount Olympus to serve as Jupiter's cupbearer.

In the Far East, Altair and its two attendant stars represented the shepherd who loved the princess in the tale of the Magpies' Bridge (see the "Lore" section for Vega, May 28).

Description•Altair and Tarazed are the most easily found stars in Aquila, and binoculars are helpful in tracing out the fainter stars of this constellation.

If you are far from electric lights, the Eagle's outline may be seen with the naked eye, its wings outstretched, headed towards the east through the glowing band of the Milky Way.

An especially bright portion of the Milky Way known as the Scutum Star Cloud is located about 20 degrees to the southwest of Altair.

Altair   July 6

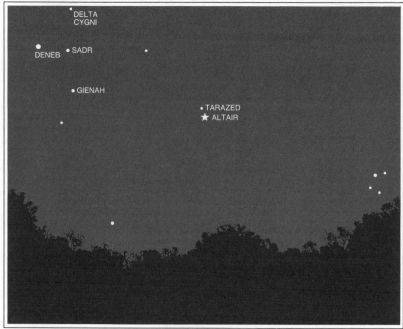

East   East-Southeast   Southeast

Altair	0.8 magnitude	2.3 absolute
al-TAIR	blue-white A7	magnitude
Alpha, α	main sequence V	19h 51m +08° 52′
Aquilae, page 214	16 light-years	September 4 transit
	+0.537 +0.387 −26	STR 1135

Location•First-magnitude Altair is 2 degrees southeast of Tarazed and 3 degrees northwest of fourth-magnitude Alshain, which may be seen with clear observing conditions. These three stars form the most distinctive feature of the constellation Aquila.

Meaning•The name Altair comes from an Arabian description of the Flying Eagle.

Lore•Altair marks the southern corner of the Summer Triangle, an asterism that may be easily found and that serves as a helpful reference. After you locate the Summer Triangle, it helps you become oriented to the positions of other stars and constellations visible in the evening sky at this time of the year.

The three stars Tarazed, Altair, and Alshain form a group some-times called the Family of Aquila. In China, these stars were called the River Drum, while in India they were known as the Ear or as the Sacred Fig Tree.

Description • Altair is a blue-white main-sequence star that shines in the sky with an apparent magnitude of 0.8. It has a surface tempera-ture of about 8,000°K and a luminosity ten times greater than that of the sun.

Main-sequence stars of Altair's luminosity are estimated to have masses about 1.7 times that of the sun. These stars have main-se-quence life expectancies in the order of nearly 2 billion years.

The spectrum of Altair features absorption lines of hydrogen and ionized metals such as magnesium, iron, and titanium.

Dschubba

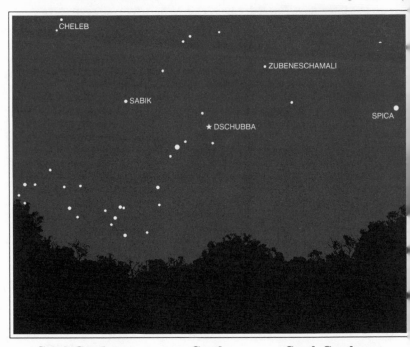

South-Southeast	South	South-Southwest
Dschubba JUBB-ah Delta, δ Scorpii, page 198	2.3 magnitude blue-white B0 main sequence V 600 light-years −0.010 −0.025 −7	−4.0 absolute magnitude 16h 00m −22° 37′ July 7 transit STR 1140

Location•Dschubba is the center star of three at the head of Scorpius, the Scorpion. It is located 3 degrees to the south of Graffias, and with a magnitude of 2.3, it is the brightest star in the trio.

Meaning•The name Dschubba is probably derived from a description of the Brow of the Scorpion.

Lore•It is said that in ancient Babylon, the three stars Graffias, Dschubba, and Pi Scorpii represented a Tree in the Garden of Light.

Description•Dschubba is one of the bright members of the group known as the Scorpius-Centaurus star association. It lies at a distance of about 600 light-years from earth and has an apparent magnitude of

2.3. The surface of this star's photosphere has a temperature of about 28,000°K, which causes Dschubba to have a blue-white color.

Dschubba's luminosity has been calculated to be 3,500 times that of the sun. Main-sequence stars of this intrinsic brightness are estimated to have masses about 7 times greater than the sun. The main-sequence life expectancy of such stars is in the order of approximately 25 million years.

The spectrum of Dschubba contains absorption lines of neutral helium and hydrogen and also reveals the presence of a close companion star with a period of about 20 days.

The technique of speckle interferometry reveals an additional companion, and observations of occultations indicate a total of four stars in the Dschubba system. With a magnitude of 3.0, the primary star is the only component that shines with a constant light; the other three stars are somewhat variable. The separation between Dschubba's components is less than 0.2 arcsecond, and the fainter stars have magnitudes of 3.3, 4.9, and 5.0.

Dschubba is one of the stars whose Sky Screen date is the same as the date of its meridian transit. These stars have such southerly declinations that they are introduced on the meridian, when they are at their maximum distance from the horizon. Even when these southerly stars are on the meridian, most observers will see them lower than the standard Sky Screen altitude above the horizon.

Antares

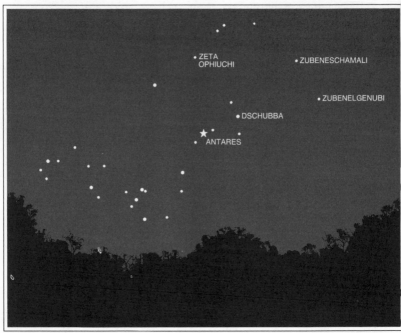

South-Southeast	South	South-Southwest

Antares an-TAIR-ease Alpha, α Scorpii, page 198	1.0 magnitude yellow-orange M1 supergiant Ib 400 light-years $-0.007 -0.023 -3$	-4.7 absolute magnitude 16h 29m $-26°$ 26′ July 15 transit STR 1169

Location•Antares is the brightest star in the constellation Scorpius. It lies 7 degrees to the southeast of Dschubba in the figure of the Scorpion.

Meaning•The name Antares comes from a Greek description "anti-Ares," the rival of Ares, which was inspired by a comparison of the star's appearance with the rusty color of the planet Mars. Mars was the Roman name for Ares, the ancient Greek god of war.

Lore•Heart of the Scorpion is a popular description of Antares, derived from this star's brightness and its location within the constellation figure.

Antares was known in ancient times as one of the four Royal Stars, along with Aldebaran, Regulus, and Fomalhaut. These first-magnitude stars are in the general vicinity of the ecliptic and served to divide the sky according to each of the four seasons. Observers in the continental United States always see Antares at a low elevation, and this results in considerable scattering of shorter-wavelength light from the star. Scattering favors the passage of light from the red end of the spectrum, and this enhances the perception of Antares as a ruddy-colored star.

In ancient China, Antares was described as the Fire Star.

History•A fifth-magnitude blue-white companion star was discovered several arcseconds from Antares in April of 1819 by the Viennese astronomer John T. Burg during an occultation by the moon.

In 1970, Antares was one of the first stars found to produce radio emissions. Astronomers at the National Radio Astronomy Observatory in Green Bank, West Virginia, detected weak radio energy coming from its atmosphere.

Description•Antares is the eleventh-brightest star visible from the midnorthern latitudes, and it shines with an average magnitude of 1.0.

This is one of the classic examples of a red supergiant star. Most of the radiant energy produced by Antares is in the invisible, infrared portion of the electromagnetic spectrum. Its most intense visible color is red, but the blend of all visible light released at this star's surface gives Antares an apparent color of yellow-orange.

This lantern of the summer night is a semiregular pulsating variable star, which fluctuates between magnitudes 0.9 and 1.8 during a period of approximately 4 years and 9 months. At its average brightness, Antares is about 6,300 times more luminous than the sun. The spectrum of Antares indicates an average surface temperature of about 3,400°K and a distance from earth of 400 light-years.

Antares is the brightest star in the Scorpius-Centaurus association, an assembly of stars that is sometimes called the Local Star Cloud.

Nearby Feature

M4: GLOBULAR CLUSTER (16h 22m −26° 27′)

The seventh-magnitude globular cluster M4 lies just over 1 degree to the west of Antares. On a clear night, binoculars show it as a hazy spot, although telescopes 12 inches in diameter or larger are needed to resolve the object into stars. At a distance of about 15,000 light-years, M4 is one of the closer globular clusters to earth.

See Scorpius constellation map, page 198.

Tau Scorpii   July 17

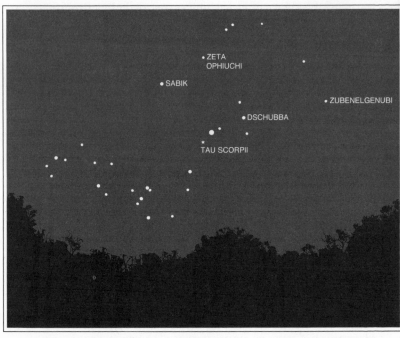

South-Southeast	South	South-Southwest
Tau Scorpii TAW SKOR-pe- eye Tau, τ Scorpii, page 198	2.8 magnitude blue-white B0 main sequence V 300 light-years −0.008 −0.025 +2	−2.8 absolute magnitude 16h 36m −28° 13′ July 17 transit STR 1176

Location·Tau Scorpii is a third-magnitude star located 2 degrees to the southeast of Antares, along the figure of the Scorpion.

Description·This blue-white main-sequence star has an apparent magnitude of 2.8 and a distance estimated to be 300 light-years from earth. Tau Scorpii's surface temperature is about 28,000°K.

Tau Scorpii's luminosity is estimated to be 1,100 times that of the sun. Main-sequence stars of this intrinsic brightness are believed to have masses about 5 times greater than that of the sun. It is believed that stars of this size spend in the order of about 50 million years as main-sequence objects.

Nearby Feature

RHO OPHIUCHI (16h 26m −23° 27′)

The sixth-magnitude double star Rho Ophiuchi is located about 3 degrees to the north of Antares. It consists of two blue-white stars, one a subgiant and the other a main-sequence object, with a separation of 3.2 arcseconds.

This double star is embedded in a reflection nebula, designated IC4604, which contains dark patches indicative of unilluminated dust clouds along our line of sight to the nebula. Immediately to the west of Rho Ophiuchi is a region of dust that almost completely hides the stars behind it. This patch has been identified as the "hole in the heavens" described by William Herschel during one of his star surveys of the 1780s. It is now known to be one of the regions of the Milky Way where new stars are currently being formed from interstellar gas and dust. It has been estimated that the dark nebula near Rho Ophiuchi contains about 1,000 solar masses of molecular hydrogen gas, as well as large quantities of interstellar dust that obscure a very young cluster of about twenty stars located deep within a molecular cloud.

Dust particles in the Rho Ophiuchi nebula and others like it are probably composed of frozen hydrogen and various nonmetallic substances. The size of these particles corresponds to that of fine clay, about 0.0005 centimeters in diameter, and they are distributed at an incredibly low average density of about fifty particles per cubic kilometer. Yet, with a diameter of one light-year, a cloud having the size of the nebula near Rho Ophiuchi contains sufficient material to block the light of stars in the background.

Since the late 1960s, astronomers have discovered molecules of increasing complexity inside interstellar dust clouds as a result of observations made using radio telescopes. Within these dust clouds, fragile molecules are protected from high-energy ultraviolet light, which would split the molecules into separate atoms, thereby destroying them.

Water, carbon monoxide, ammonia, formaldehyde, and ethyl alcohol are some of the many organic chemicals that have been discovered. The presence of these chemicals, which on earth are associated with the biological activity of living things, is one of the major astronomical discoveries of recent decades.

British astrophysicist Fred Hoyle and his colleague Wickramasinghe, in an experiment to study the nature of the interstellar medium, showed that certain optical properties of interstellar dust grains, observed in various parts of the Milky Way, closely match those of a mixture of spore-producing bacteria and microscopic carbon spheres produced when bacteria are burned. Hoyle therefore feels that the Galaxy may contain vast quantities of living material.

See the constellation maps for Ophiuchus (page 178), Scorpius (page 198), and Orion (page 332).

Epsilon Scorpii   July 20

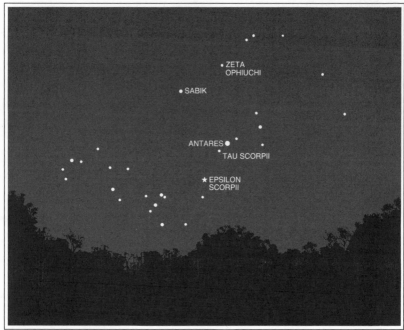

ZETA OPHIUCHI

SABIK

ANTARES
TAU SCORPII

★ EPSILON SCORPII

South-Southeast	South	South-Southwest
Epsilon Scorpii	2.3 magnitude	+0.7 absolute
EP-sigh-lon	yellow K2	magnitude
SKOR-pe-eye	giant-subgiant	16h 50m −34° 18′
Epsilon, ϵ	III–IV	July 20 transit
Scorpii, page 198	70 light-years	STR 1190
	−0.610 −0.255 −3	

Location•Second-magnitude Epsilon Scorpii is located 6 degrees to the southeast of Tau Scorpii, along the figure of the Scorpion.

Lore•In the traditional image of Scorpius, this star marks the beginning of the Scorpion's tail.

Description•Epsilon Scorpii is a yellow giant to subgiant with an apparent magnitude of 2.3. Its surface temperature is about 4,300°K, and its luminosity is 45 times greater than that of the sun.

Epsilon Scorpii's spectrum shows strong lines of neutral metals. Calcium, for example, displays an intense absorption line at a wavelength of 4,227 angstrom units. Epsilon has yearly proper-motion components of 0.610 arcsecond towards the west (decreasing right

ascension) and 0.255 arcsecond towards the south (decreasing declination). The combined rate of proper motion is 0.661 arcsecond per
year towards the east-southeast on the celestial sphere. Examination
of this star's distance from earth and motion shows marked differences from those of other bright stars in Scorpius. This indicates that
Epsilon is unrelated to the Scorpius-Centaurus association. Epsilon
Scorpii's rate of proper motion after 2,800 years would move the star
across the celestial sphere a distance equal to the apparent diameter of
the full moon.

Epsilon is 70 light-years from earth, and it is one of only a few
bright stars in its constellation that does not belong to the Scorpius-
Centaurus association of stars.

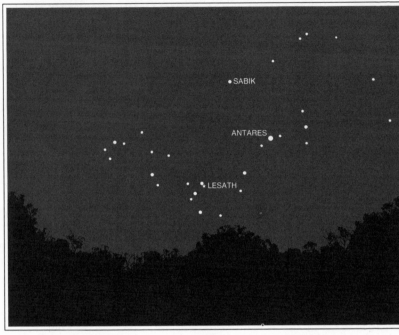

South-Southeast   **South**   **South-Southwest**

Lesath LESS-ath Upsilon, υ Scorpii, page 198	2.7 magnitude blue-white B2 subgiant IV 450 light-years −0.001 −0.032 +8	−3.0 absolute magnitude 17h 31m −37° 18′ July 30 transit STR 1231

Location•Lesath is a third-magnitude object, and it is one of two close stars that mark the tip of the Scorpion's tail. Lesath is located 16 degrees to the southeast of Antares.

Meaning•The name Lesath is based on a Latin version of an Arabic description of a foggy spot at the tip of the tail of Scorpius. It was probably first used by the French astronomical writer Joseph Scaliger around the year 1600. A blend of light from Lesath and its very close neighbor Shaula may at times resemble a cloudy spot rather than two distinct stars.

Description•Lesath is a blue-white subgiant star with an apparent magnitude of 2.7 and a distance from earth of 450 light-years. This star has a surface temperature of 23,000°K and an intrinsic brightness about 1,300 times greater than the sun's.

Lesath's spectrum is characterized by lines of neutral helium, ionized silicon, oxygen, and magnesium. Hydrogen lines are also evident. Various absorption lines, as those of hydrogen, helium, and oxygen, are used as luminosity criteria in spectral class B1 stars such as Lesath. The annual proper motion, east 0.001 arcsecond and south 0.029 arcsecond, corresponds with that of other stars in Scorpius including Antares and Shaula. This motion indicates membership in the Scorpius-Centaurus association.

M7: OPEN STAR CLUSTER (17h 53m −34° 48′)

This is another open star cluster that is easy to find and visible with binoculars. M7 is located 5 degrees to the northeast of Lesath. On a clear, dark night, it looks like a small cloud to the naked eye and covers an area nearly 1 degree in diameter. The distance to M7 is about 800 light-years, and its age is estimated to be in the order of 250 million years.

M7 can be seen on the Scorpius constellation map, page 198.

Shaula

Shaula   July 31

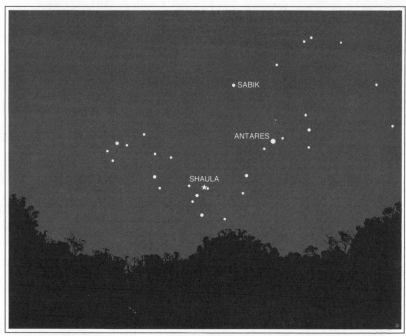

South-Southeast	South	South-Southwest
Shaula	1.6 magnitude	−3.4 absolute
SHAW-la	blue-white B1	magnitude
Lamda, λ	subgiant IV	17h 34m −37° 06′
Scorpii, page 198	350 light-years	July 31 transit
	+0.001 −0.029 −3	STR 1234

Location•Shaula is a second-magnitude star located at the tip of the tail of Scorpius, adjacent to Lesath. The separation between these two stars is approximately equal to the diameter of the full moon.

Meaning•The name is derived from an ancient description of the Stinger of the Scorpion.

Lore•Shaula and the other stars that form the tail of Scorpius are said to have also marked a Polynesian asterism known as the Fish Hook of Maui.

Description•This is a variable star of the Beta Cephei variety and has an average apparent magnitude of 1.6. Shaula is a blue-white subgiant, as is characteristic of this class of variable star. The magnitude ranges from about 1.59 to 1.65 over a period of 0.214 day. Shaula is the brightest-known example of this type of variable, which is believed to be at the start of its evolution away from the main sequence.

The distance to Shaula is approximately 350 light-years, and it has an average luminosity about 1,900 times that of the sun.

Nearby Feature

M6: OPEN STAR CLUSTER (17h 39m −32° 11′)

The open star cluster designated number six on Messier's list of nonstellar objects is located 5 degrees of arc to the north of Shaula. This cluster contains several hundred stars and is located about 1,300 light-years from the earth. It is easily visible with binoculars.

Open clusters contain stars that were formed from common Milky Way clouds at approximately the same time. Clusters present an array of stars of different masses and luminosities yet with approximately the same composition and age. Spectral studies, for example, can provide a "snapshot" of the evolution of stars within a particular cluster, giving a wide variety of information about the processes involved and the characteristics of stars of different temperatures and luminosities.

M6 can be seen on the Scorpius constellation map, page 198.

Star clusters are also very useful for estimating distances in space. The H-R diagram of a particular open cluster may be compared with that of the Hyades cluster in the constellation Taurus, whose distance is well known. As a result of this comparison, the absolute magnitudes of various cluster stars may be estimated. A comparison of absolute and apparent magnitudes for these cluster stars gives an estimate of the open cluster's distance from earth.

The H-R diagram of a star cluster also gives an indication of the cluster's age. Astronomers observe the cluster's main sequence to see what spectral class of stars has begun to evolve into giants. Theoretical studies of stellar evolution then provide an age estimate. The upper-left endpoint of the main sequence of M6, for example, indicates that this open star cluster formed about 100 million years ago.

The Bright Stars of Summer

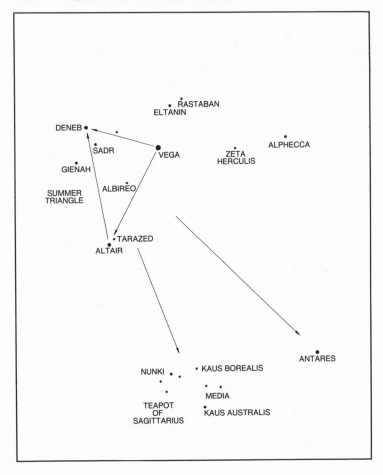

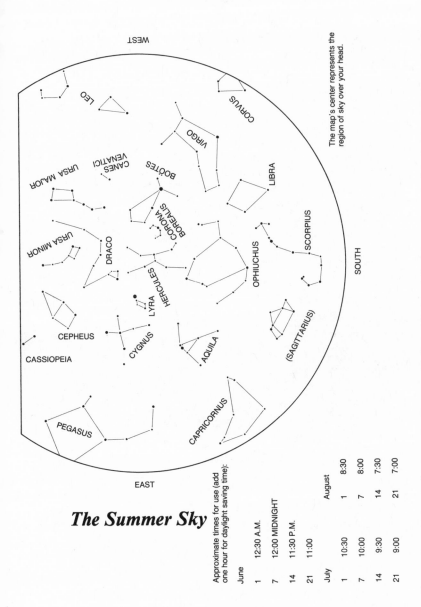

WEST

The map's center represents the region of sky over your head.

LEO

CORVUS

VIRGO

CANES VENATICI

URSA MAJOR

BOÖTES

LIBRA

CORONA BOREALIS

SCORPIUS

URSA MINOR

DRACO

SOUTH

OPHIUCHUS

LYRA

HERCULES

CEPHEUS

CYGNUS

(SAGITTARIUS)

CASSIOPEIA

AQUILA

PEGASUS

CAPRICORNUS

EAST

The Summer Sky

Approximate times for use (add one hour for daylight saving time):

June	
1	12:30 A.M.
7	12:00 MIDNIGHT
14	11:30 P.M.
21	11:00

July	
1	10:30
7	10:00
14	9:30
21	9:00

August	
1	8:30
7	8:00
14	7:30
21	7:00

CALENDAR OF
August Stars

Sargas   August 1

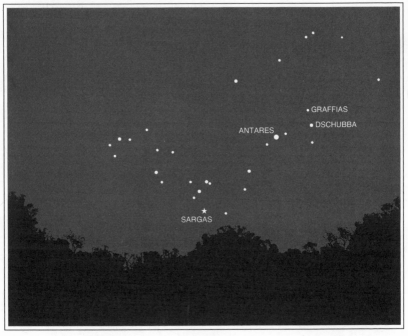

South-Southeast	South	South-Southwest
Sargas SAR-gas Theta, θ Scorpii, page 198	1.9 magnitude blue-white F0 supergiant Ib 500 light-years +0.016 +0.000 +1	−4.5 absolute magnitude 17h 37m −43° 00′ August 1 transit STR 1237

Location•Second-magnitude Sargas, in the tail of Scorpius, may be found about 5 degrees to the south of Lesath and Shaula.

Meaning•Sargas is believed to have been the name of a Mesopotamian star god.

Description•Sargas has the most southerly declination of all the bright stars featured in this guide. Because of this, observers need an unobstructed view of the southern sky in order to see this star.

Sargas is a blue-white supergiant with an apparent magnitude of 1.9 and a distance from earth of 500 light-years. Its surface temperature is about 7,000°K, and the luminosity is equal to the brightness of 5,500 suns.

Spectra of F0 stars such as Sargas contain strong absorption lines of hydrogen and lines of ionized metals including titanium, iron, magnesium, and calcium. The hydrogen lines are of nearly equal intensity and width in all F0 luminosity categories from main sequence through subgiant; therefore, these lines are of little use as criteria of luminosity for F0 stars. On the other hand, lines of ionized titanium intensify with increasing luminosity and were used by astronomers to determine that Sargas is in the supergiant luminosity category.

Kappa Scorpii August 2

South-Southeast	South	South-Southwest
Kappa Scorpii kap-ah SKOR-pe-eye Kappa, κ Scorpii, page 198	**2.4 magnitude blue-white B2 subgiant IV 450 light-years −0.007 −0.029 −14**	**−3.3 absolute magnitude 17h 42m −39° 02′ August 2 transit STR 1242**

Location•Second-magnitude Kappa Scorpii, near the end of the Scorpion's tail, is the last of our featured stars in Scorpius to reach the meridian. It is found about 2 degrees to the southeast of Lesath and Shaula.

Description•Kappa Scorpii is a variable star of the Beta Cephei variety. These stars have exhausted the supply of hydrogen in their cores as a result of its conversion into helium. Subgiant stars of this type undergo some instability as the process of hydrogen fusion develops in shells surrounding their cores and the stars' diameters slowly increase.

Kappa is a blue-white subgiant with an average apparent magnitude of about 2.4. Its distance from earth is about 450 light-years.

Kappa Scorpii has a luminosity of 1,700 suns and a surface tempera-ture of about 23,000°K. The magnitude varies from 2.39 to 2.42 during a period of 4 hours and 47.8 minutes. This range of magnitude is equivalent to a brightness variation of about 3 percent.

Hydrogen and neutral helium are the strongest absorption lines in the spectrum of Kappa Scorpii. These lines are of intermediate inten-sity and width in subgiants of this spectral class. Since the hydrogen and helium lines decrease in intensity with increasing luminosity from main sequence to supergiant in class B2 stars, they serve as indicators of the category of luminosity to which this type of star belongs.

The proper motion of Kappa is similar to that of Tau Scorpii, Antares, and other members of the Scorpius-Centaurus association, indicating its membership in this assembly of stars.

The distance between Kappa and the sun is decreasing at the rate of 14 kilometers per second.

Enif   August 3

| East-Northeast | East | East-Southeast |

| Enif
ENN-if
Epsilon, ϵ
Pegasi, page 238 | 2.4 magnitude
yellow K2
supergiant Ib
800 light-years
+0.030 +0.005 +5 | −4.6 absolute
magnitude
21h 44m +09° 53′
October 2 transit
STR 1245 |

Location·Enif is the first of our featured stars in the constellation Pegasus to appear on the Sky Screen. This second-magnitude star is located 25 degrees to the east of Altair.

Meaning·The name Enif is believed to have come from a description of this star's location at the nose or lip of the Flying Horse, Pegasus.

Lore·Traditional portrayals of Pegasus usually show Enif near the end of the horse's nose.

Description·The apparent magnitude of Enif is 2.4 and it is a yellow supergiant. Located about 800 light-years from earth, it has a luminosity 6,000 times greater than the sun's. The star's surface temperature is about 4,000°K.

Nearby Feature

M15: GLOBULAR STAR CLUSTER (21h 29m +12° 05′)

M15 is one of the most attractive globular star clusters visible in northern skies. This cluster probably contains over 250,000 stars, and its distance is approximately 34,000 light-years. The apparent magnitude of M15 is about 6.5, and through binoculars it resembles a single hazy star. A telescope of at least an 8-inch diameter at 100 power is usually needed to begin to resolve individual stars around the fringes of M15.

This globular cluster is located 4 degrees to the northwest of the star Enif.

See the following constellation map.

Pegasus, *the Flying Horse*

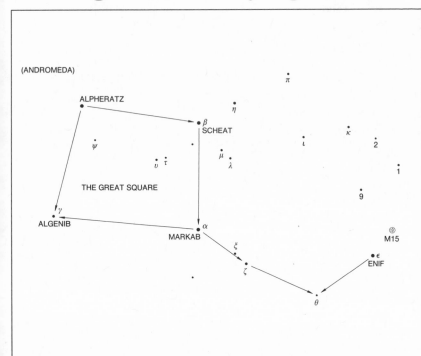

BRIGHT STARS OF PEGASUS

August 3	Enif	2.4	yellow	supergiant
August 11	Scheat	2.5	yellow-orange	giant
August 20	Markab	2.5	blue-white	giant
September 6	Algenib	2.8	blue-white	subgiant

PEG-uh-suss
Peg
Pegasi

Evening Season•September – January

Sky Track•Pegasus rises in the northeastern sky, passes overhead near the zenith, and sets towards the northwest.

Lore•The image of a flying horse has been found on numerous artifacts from Mesopotamia and other parts of the ancient world. In much of this region, the stars of Pegasus came to represent a winged horse.

This constellation was identified in ancient Israel as the horse of Nimrod, the great hunter mentioned in Genesis 10:8.

In Greek mythology, Pegasus was the offspring of Neptune and Medusa, created at his father's command from a mixture of sea foam and the blood that fell when Perseus severed Medusa's head.

The Flying Horse arrived at Mount Helicon, the mountain of the Muses, where his hoof opened a fissure in the soil from which the waters of the Hippocrene spring poured forth. At a later time, Pegasus formed the Pirene spring of Corinth in a similar way.

The legends tell us that there lived in Corinth a handsome young man named Bellerophon, who longed to ride upon the wonderful flying horse. Athena answered his prayers by providing a golden bridle that charmed Pegasus into accepting Bellerophon as his rider.

The pair went on to many adventures together, including the conquest of a dreadful creature known as the Chimera, which possessed the combined body parts of a lion, a serpent, and a goat.

Bellerophon's pride and self-confidence grew until he felt entitled to a place alongside the gods on Olympus. However, Pegasus knew that such hubris was despised by the gods, and when Bellerophon urged him to fly to Olympus, Pegasus threw Bellerophon to the earth, where he was forced to wander alone for the rest of his life.

Pegasus, however, was welcomed on Mount Olympus, where he remained as the Winged Horse of Zeus and the bearer of divine thunderbolts.

Description·The constellation Pegasus is dominated by the asterism of the Great Square. This represents the body of Pegasus, who is portrayed upside down in the sky, probably due to effects of precession.

The Great Square forms a centerpiece in the autumn evening sky that serves as a base from which to search out neighboring bright stars.

Caph

August 7

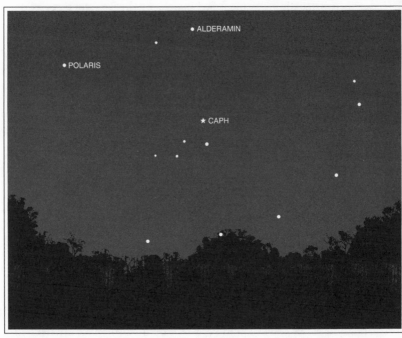

North-Northeast	Northeast	East-Northeast
Caph Kaff Beta, β Cassiopeiae, page 242	2.3 magnitude blue-white F2 subgiant IV 46 light-years +0.526 −0.177 +11	+1.5 absolute magnitude 00h 09m +59° 09′ November 8 transit STR 1260

Location•We turn to the northern part of the sky in order to locate Caph, a second-magnitude star in the W-shaped asterism of the constellation Cassiopeia. Caph marks the upper-right corner of the W's figure. This star is located 21 degrees from Alderamin and 30 degrees from Polaris.

Meaning•The name Caph was taken from an Arabian title for bright stars in Cassiopeia, which meant "the extended right hand of the Pleiades."

Lore•The W-shaped asterism of Cassiopeia is also seen as an M, depending on the constellation's orientation as it moves around the North Celestial Pole.

Description•Caph is the brightest example of a rare class of pulsating variable stars of which Delta Scuti is the prototype. These subgiant stars are in the process of evolving away from the main sequence.

Caph fluctuates between magnitudes 2.25 and 2.31 during a short period of only 2 hours and 30 minutes, representing a brightness change of about 6 percent.

At its average magnitude, Caph is 20 times brighter than the sun and has a surface temperature of about 7,000°K. Its distance from earth has been determined by the method of trigonometric parallax to be 46 light-years, and it shines with a blue-white color.

Spectral studies reveal a close companion with an orbital period of about 27 days.

Nearby Feature

TYCHO'S STAR: AN HISTORIC SUPERNOVA (00h 25m +63° 50′)

Although it was visible for only a few years and faded from view before the invention of the telescope, Tycho's star is remembered as one of the brightest supernovae in history, as well as one of the last. When it first blazed into view, it rivaled the brilliance of the planet Venus and was a spectacular addition to the sky.

One of the first to marvel at this "new" star's glory was the young Danish nobleman Tycho Brahe. As a boy, Tycho Brahe had been fascinated by the stars and planets. He continued to satisfy this enthusiasm while studying at the University of Copenhagen, even though his family encouraged him to prepare for a political career.

The great supernova appeared in November of 1572 and Tycho soon recognized its special nature. He compared its brightness to that of Venus and other planets and carefully recorded its location. The object's position relative to other stars in Cassiopeia remained fixed, indicating that it probably belonged to the realm of the stars. Tycho summarized his observations and conclusions in a book entitled *De Nova Stella,* which presented the first evidence of activity in the supposedly immutable space of the stars. This work also introduced the term *nova* as a description for a star that suddenly bursts into view.

Tycho's star of 1572 was extremely massive, in the throes of a catastrophic explosion that climaxed its evolution. The small minority of stars that retain more than about 1.4 times the sun's mass after their turn as red giants possess too much gravitational energy, and their demise is marked by a supernova explosion rather than by formation of a planetary nebula. In such a cataclysm, the outer layers of the star are blasted away by a sudden burst of energy, and the star's core is exposed.

Supernovae may shine with one billion times the luminosity of the sun and may be observed with the naked eye at distances of thousands of light-years.

Cassiopeia, *the Queen*

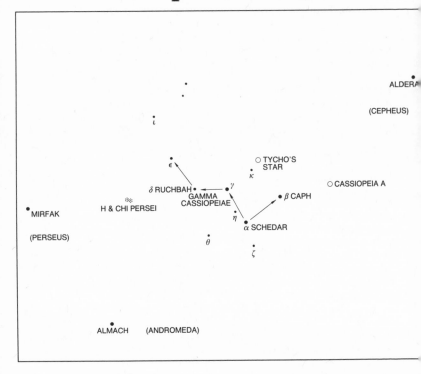

BRIGHT STARS OF CASSIOPEIA

August 7	Caph	2.3	blue-white	subgiant
August 17	Schedar	2.2	yellow	giant
August 18	Gamma Cassiopeiae	2.6	blue-white	subgiant
August 25	Ruchbah	2.7	blue-white	giant-subgiant

KASS-ee-oh-PEE-ah
Cas
Cassiopeiae

Evening Season•August – March

Sky Track•Cassiopeia is a north circumpolar constellation, and it is above the horizon at all times throughout the year. It lies on the opposite side of the North Star from the handle of the Big Dipper.

Lore•The stars of Cassiopeia are said to represent a Queen of Ethiopia who, in Greek mythology, offended the Nereids, or sea nymphs, by her vain boasting. It was said that the sea nymphs caused Cassiopeia to be placed in the northern sky, bound to a chair so that as she revolved around the celestial pole she would at times assume a humbling upside-down position. For this reason, Cassiopeia is often described as the Lady in the Chair.

In order to visualize Cassiopeia's Chair, the fourth-magnitude star Kappa Cassiopeiae is added to the W to form the front of the chair's seat.

Description·The W of Cassiopeia is one of the most familiar asterisms in the sky, and it forms an apparent counterpoise to the Big Dipper, with the North Star forming the pivot point, as these wonderful figures circle the northern sky in a counterclockwise direction.

Cassiopeia lies in a region of the Milky Way that appears relatively narrow to the naked eye because it is opposite the Galactic center in Sagittarius, yet the Cassiopeia region is a lovely sight through binoculars or a wide-angle, rich-field telescope.

Scheat

Northeast **East-Northeast** **East**

Scheat	2.5 magnitude	−1.4 absolute
SHEE-at	yellow-orange M2	magnitude
Beta, β	giant II–III	23h 04m +28° 05′
Pegasi, page 238	200 light-years	October 23 transit
	+0.188 +0.142 +9	STR 1276

Location•Third-magnitude Scheat is located at the northwestern corner of the Great Square of Pegasus. Its position is about 20 degrees to the northeast of the star Enif.

Meaning•The name Scheat appears to have originated as part of a description of the star Delta in the constellation Aquarius. The name was then later applied to this star, Beta Pegasi.

Lore•Scheat's entry into the evening sky marks the arrival of the Great Square of Pegasus. Although the weather may still be quite hot, Scheat reminds us that cooler days are not many weeks away.

This star represented the shoulder of Pegasus in Ptolemy's description of that constellation.

Description•When you see the Great Square of Pegasus in the eastern sky, it is tilted so that Scheat ascends first, as if it were at the top of a diamond.

Scheat is a yellow-orange variable star with a magnitude range from 2.3 to 2.7 and an average magnitude of 2.5. Its luminosity varies from about 250 to 350 times the brightness of the sun, and the average surface temperature is about 3,300°K.

The apparent diameter of Scheat was found to be 0.021 arcsecond, which represents an actual diameter nearly the size of earth's orbit.

Nearby Feature

PERSEID METEOR SHOWER

Every year on the night of August 11, the Perseid meteor shower reaches its maximum frequency. At the present time this is one of the most spectacular and predictable of all meteor displays. The Perseids were first mentioned in Chinese records of the year A.D. 36, and they are believed to result from earth's collision with fragments of the comet Swift-Tuttle that separated and were distributed along the comet's solar orbit. These meteors may be seen when earth reaches the vicinity of Swift-Tuttle's orbit in August and begins to sweep up debris from this comet.

Perseid meteors trace bright streaks in the sky that appear to radiate from near the constellation Perseus, hence their name. The majority of these meteors are formed by sand-grain-sized particles of rock, which enter our atmosphere and begin to glow from frictional heating in the air.

Perseids most frequently are of the second magnitude in brightness, and about one meteor per minute may be seen during the hours near a shower's maximum.

The name *meteoroid* is applied to the object while it is in space, before it enters the atmosphere, is heated, starts to vaporize, and becomes visible. *Meteor* describes the object and its envelope of glowing gas as it plunges through the atmosphere. The envelope of incandescent gas may be up to several hundred meters in diameter, and meteors are usually visible at altitudes from as high as 120 kilometers down to about 80 kilometers, where the increased density of air causes the vast majority of meteors to disintegrate completely. However, if the object is large enough, fragments may survive and reach the ground, where the pieces are called *meteorites.*

Media   August 12

South-Southeast   **South**   **South-Southwest**

Media	2.7 magnitude	+1.1 absolute
ME-dee-ah	yellow K3	magnitude
Delta, δ	giant III	18h 21m −29° 50′
Sagittarii, page 248	70 light-years	August 12
	+0.039 −0.029 −20	STR 1281

Location•Media marks the point where the top of the "spout" joins the rest of the teapot-shaped asterism in Sagittarius. This third-magnitude star is featured on the sky screen as it crosses the meridian low in the southern sky. It is located about 12 degrees to the northeast of Shaula and Lesath.

Meaning•The name Media was adapted from the Latin description of this star's location at the middle part of the Archer's bow.

Lore•In classical sky lore, Media marked the point where the arrow crossed the middle of the Archer's bow.

History•A thirteenth-magnitude companion, at a distance of 58 arc-seconds, was discovered by the American astronomer Thomas Jefferson Jackson See in 1896.

Description•Media is a yellow giant that appears to have a magnitude of 2.7. The trigonometric parallax of this star indicates a distance of 70 light-years, which means that Media is 30 times more luminous than the sun in order for it to shine with the magnitude observed on earth. Its surface temperature is estimated to be 4,100°K.

This and other stars of class K3 show prominent lines of neutral metals in their spectra. Ionized metals and titanium oxide molecules are also characteristic of stars of this class. The proper motion of Media is 0.039 arcsecond per year eastward and 0.029 arcsecond per year southward on the celestial sphere. Every second the star Media moves 20 kilometers closer to the sun.

Sagittarius, *the Archer*

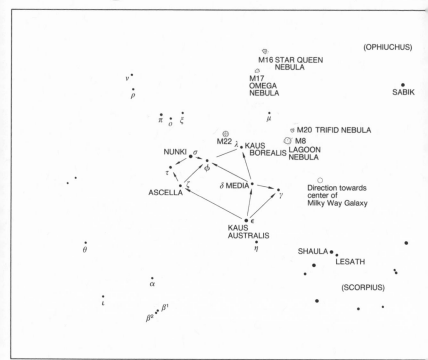

BRIGHT STARS OF SAGITTARIUS

August 12	Media	2.7	yellow	giant
August 13	Kaus Australis	1.8	blue-white	subgiant
August 14	Kaus Borealis	2.8	yellow	giant
August 21	Nunki	2.1	blue-white	main sequence
August 23	Ascella	2.6	blue-white	giant and subgiant

saj-ih-TAY-rih-us
Sag
Sagittarii

Evening Season•August

Sky Track•Due to the degree of its southern declination, observers in the continental United States see Sagittarius move from southeast to southwest, never reaching far from the horizon.

Lore•This constellation was known to ancient Greeks as the Archer, the most popular classical description of Sagittarius.

Eratosthenes, in the third century B.C., described the figure as a Satyr, with the combined body of a man and horse. Sometimes this Satyr has been identified as Chiron, the wise and gentle teacher of Asclepias, Achilles, and Hercules. Usually, however, Chiron is associated with the constellation Centaurus, the Satyr of the South, and Sagittarius is shown as a fearsome creature. In all representations, the Archer-Centaur's arrow is aimed towards the west, at the Scorpion. The stars Kaus Borealis, Media, and Kaus Australis mark the bow, while Gamma Sagittarii represents the arrow tip.

The figure was sometimes known simply as the Bow, a title used by both Cicero and Germanicus Caesar, a Roman general and grandson of Augustus. Both translated work of the Greek astronomical poet Aratus into Latin and thereby helped to preserve Hellenic sky myths.

An asterism known as the Teapot of Sagittarius is very distinctive and provides the easiest way to identify this constellation. Third-magnitude star Kaus Borealis marks the top of the Teapot's lid, and Gamma Sagittarii, also of third magnitude, is at the tip of the figure's spout. Second-magnitude Nunki is at the top of the Teapot's handle.

Another asterism, called the Milk Dipper, is formed from stars in the Teapot's handle and lid, plus fourth-magnitude Mu Sagittarii, found 5 degrees to the north of the lid.

Description• The direction of the center of the Milky Way Galaxy lies 5 degrees to the west of the Teapot's spout, and as a result, the constellation Sagittarius is a treasury of star clusters, bright and dark nebulae, and other types of Galactic objects.

On a clear night, far from artificial lights, the glowing band of the Milky Way appears to rise as steam from the Teapot's spout.

Kaus Australis   August 13

South-Southeast   **South**   **South-Southwest**

Kaus Australis	1.8 magnitude	−1.5 absolute
koss oss-TRAY-lis	blue-white B9	magnitude
Epsilon, ε	subgiant IV	18h 24m −34° 23m
Sagittarii, page 248	150 light-years	August 13 transit
	−0.032 −0.125 −15	STR 1284

Location· Second-magnitude Kaus Australis is the brightest star in Sagittarius and is located at the base of the Teapot's spout, 5 degrees south of Media.

Meaning· The proper name Kaus Australis, meaning "the southern star in the bow," was introduced by the Sicilian astronomer Giuseppi Piazzi at the beginning of the nineteenth century. None of the bow stars had names handed down from antiquity, and Piazzi combined Latin and Arabic words as he invented these designations. *Kaus* came from an Arabic name for bow, while *Borealis, Meridionalis (Media),* and *Australis* described the northern, middle, and southern stars in this figure.

History•Although Kaus Australis is the brightest star in the constellation, it is designated Epsilon Sagittarii. An examination of the brightnesses and positions of the stars Alpha, Beta, Gamma, Delta, and Epsilon Sagittarii provides an example of the fact that Bayer's Greek-letter star nomenclature was sometimes based on an order other than relative brightness.

Description•Kaus Australis has an apparent magnitude of 1.8 and is a blue-white subgiant. Its luminosity is estimated to be about 350 times greater than that of the sun, and this leads to a distance estimate of about 150 light-years, based on a comparison of the star's luminosity and apparent magnitude. The surface temperature of Kaus Australis is about 10,000°K.

Binoculars are helpful in observing the color contrast between Kaus Australis and the yellow stars Media and Gamma Sagittarii, located at the top of the Teapot's spout. A seventh-magnitude blue-white optical companion to Kaus Australis may be seen with binoculars 3.3 arcminutes to the north of Kaus.

Kaus Borealis   August 14

South-Southeast	South	South-Southwest

Kaus Borealis
 koss BORE-ee-
    ALICE
 Lambda, λ
 Sagittarii, page 248
	2.8 magnitude
	 yellow K1
	 giant III
	 60 light-years
	 −0.043 −0.185 −43
		+1.4 absolute
		    magnitude
		 18h 28m −25° 25′
		 August 14 transit
		 STR 1288

Location•Kaus Borealis marks the top of the lid of the teapot-shaped asterism in the constellation Sagittarius. It is a third-magnitude star.

Meaning•The name Kaus Borealis means "the northern star in the bow." It was one of the star names devised by astronomer Giuseppi Piazzi.

Lore•In addition to representing the top to the Teapot, Kaus Borealis also marks a bend in the handle of an asterism known as the Milk Dipper. This is one of several celestial dippers in addition to the familiar ones in the northern sky. It is formed from stars in the Teapot of Sagittarius with the addition of fourth-magnitude Mu Sagittarii, which represents the end of the Milk Dipper's handle.

Description•Kaus Borealis is a yellow giant star that shines in our skies with an apparent magnitude of 2.8. Its spectrum indicates a surface temperature of about 4,400°K.

The distance to Kaus Borealis has been determined by the method of trigonometric parallax to be 60 light-years. This implies that the star is about 25 times more luminous than the sun.

Nearby Features•These three nebulae can be seen on the Sagittarius constellation map, page 248.

M8: THE LAGOON NEBULA (18h 02m −24° 23′)

This bright gaseous nebula has a diameter somewhat larger than the disk of the full moon and is located about 5 degrees to the west of Kaus Borealis. Small telescopes show M8 as a faint patch of light, associated with a small open cluster of stars. An 8-inch or larger telescope is usually required to show the dark lane that characterizes the Lagoon Nebula. Gaseous nebulae such as this glow as a result of the stimulation of hydrogen gas by high-energy ultraviolet light from nearby stars. The Lagoon Nebula has a distance of about 4,500 light-years from earth.

M17: THE HORSESHOE OR OMEGA NEBULA (18h 20m −16° 12′)

M17 is another gaseous or "emission" nebula, located about 9 degrees to the north of Kaus Borealis. This glowing gas cloud is visible with binoculars, although a 6-inch telescope, used with low power, is needed to show the horseshoe shape. This nebula has an apparent diameter about two-thirds the size of the full moon, and it lies about 3,000 light-years from us.

M16: THE STAR QUEEN NEBULA (18h 18m −13° 48′)

M16 consists of a cluster of young stars and a surrounding region of nebulosity. It was described as early as 1746 by the Swiss astronomer P. L. de Cheseaux, and the first photograph of the nebula was made by E. E. Barnard in 1895.

Robert Burnham, of the Lowell Observatory, named this the Star Queen Nebula in his *Celestial Handbook:*

> M16 is one of the most spectacular of the diffuse nebulae, and shows an astonishing amount of fascinating detail. Thrusting boldly into the heart of the cloud rises a huge pinnacle like a cosmic mountain, the celestial throne of the "Star Queen" herself, wonderfully outlined in silhouette . . .

M16 is located 12 degrees north of Kaus Borealis, in the constellation Serpens. Its open star cluster is believed to be less than one million years old, and the distance of M16 from earth is estimated to be approximately 8,000 light-years.

Schedar August 17

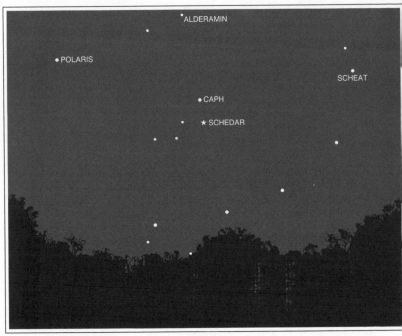

North-Northeast	Northeast	East-Northeast

Schedar SHED-are Alpha, α Cassiopeiae, page 242	2.2 magnitude yellow K0 giant II–III 150 light-years +0.053 −0.027 −4	−1.0 absolute magnitude 00h 41m +56° 32′ November 16 transit STR 1299

Location•Schedar is a second-magnitude star that marks the lower right-hand point of the W of Cassiopeia. It is 5 degrees from the star Caph.

Meaning•The name Schedar is adapted from an Arabic description of this star as the Breast of Cassiopeia, its position according to Ptolemy's star catalogue.

Lore•In one old representation of Cassiopeia, stars of the W form Cassiopeia's Chair. In this asterism Schedar marks the tip of one of the chair's back legs.

History•Several nineteenth-century observers, including John Herschel and Friedrich Argelander, reported that Schedar showed slight and unpredictable variations in brightness. In recent decades, however, no definite changes in magnitude have been observed.

Description•Schedar has an apparent magnitude of 2.2 and a distance calculated to be about 150 light-years. It is a yellow giant star with a surface temperature of approximately 4,500°K. Its luminosity is equal to the brightness of 200 suns.

There is a ninth-magnitude optical, and therefore unrelated, companion star at a distance of 64 arcseconds from Schedar. This blue-white star was discovered by William Herschel in 1781, the same year that he discovered the planet Uranus and became famous.

Nearby Feature

ETA CASSIOPEIAE: A 12-ARCSECOND DOUBLE STAR
(00h 49m +57° 49′)

Here is another example of the many double stars discovered by William Herschel, in this case in 1779. Eta Cassiopeiae has a combined magnitude of 3.4 and is located less than 2 degrees from Schedar, nearly on a line with Gamma Cassiopeiae, the center star in the W.

The components of Eta have apparent magnitudes of 3.5 and 7.5 and the colors white and yellow-orange. The separation is 12 seconds of arc, and the period of this star system is calculated to be 480 years. The dramatic difference in magnitude between components provides a lovely sight at magnifications of at least 100×. Due to the faintness of the secondary star, the pair's color contrast is not obvious. It is about 19 light-years from earth.

Gamma Cassiopeiae August 18

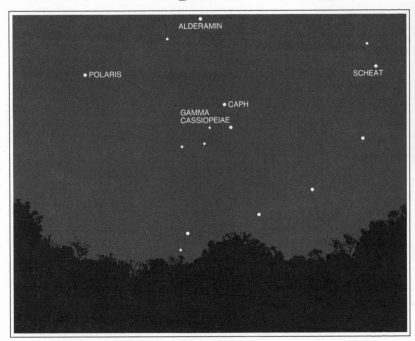

North-Northeast	Northeast	East-Northeast
Gamma Cassiopeiae GAH-ma kass-ee-oh-PEA-aye Gamma, γ Cassiopeiae, page 242	2.5 magnitude blue-white B0 subgiant IV 200–600 light years (?) +0.025 +0.000 −7	absolute magnitude uncertain 00h 57m +60° 43′ November 20 transit STR 1301

Location•Gamma Cassiopeiae is the star at the central peak of the W-shaped asterism of Cassiopeia. It is a third-magnitude object about 5 degrees from Schedar and 6 degrees from Caph.

History•This is the first star in which bright emission lines were discovered. Pietro Angelo Secchi made this observation in 1866.

Until 1910, Gamma Cassiopeiae seemed to have a constant brightness of magnitude 2.3. In that year it slowly began to brighten, reaching magnitude 1.8 in 1936, then reaching magnitude 1.6 in April 1937. By the beginning of 1938 it had faded to magnitude 2.3 and reached a minimum of magnitude 3.1 during 1940. The star slowly brightened again, attaining magnitude 2.5 in 1954 and 2.2 in 1976. In 1982 its magnitude was listed as 2.5.

Description·Gamma Cassiopeia is an erratic variable star, which is ejecting matter into space and is surrounded by an expanding shell of gas. It is a blue-white spectral type B0 subgiant, which undergoes violent and unpredictable changes in brightness, surface temperature, color, and diameter that astronomers cannot fully explain.

At the time of its maximum brightness in 1937, Gamma Cassiopeiae suddenly, during a period of a few months, dropped in surface temperature from 12,000°K to about 8,500°K. It is believed that at this time the star ejected its spherical shell. The third Small Astronomy Satellite *(SAS-3)* detected emissions of weak X rays from Gamma Cassiopeia during 1976, and there are indications the star is surrounded by a rotating disklike envelope.

The adjacent nebulae IC 59 and IC 63 seem to be associated with Gamma Cassiopeiae. They have curious dark triangular structures that may have been shaped by stellar winds blowing from Gamma.

The distance to Gamma Cassiopeiae is very difficult to determine because it is beyond the range of accurate trigonometric parallax and its abnormal characteristics defy estimate of its absolute magnitude.

An eleventh-magnitude companion star at a distance of 2.3 arcseconds was discovered by S. W. Burnham at Lick Observatory in 1888. It is very difficult to see because of the brightness difference between it and the primary.

Markab
August 20

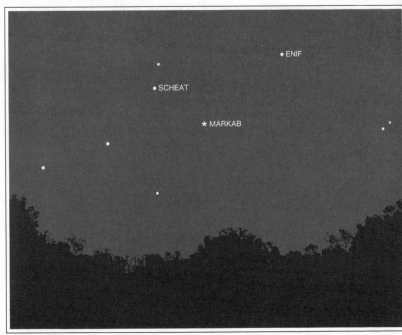

East-Northeast	East	East-Southeast
Markab MAR-kab Alpha, α Pegasii, page 238	2.5 magnitude blue-white B9.5 giant III 110 light-years +0.062 −0.038 −4	−0.1 absolute magnitude 23h 05m +15° 12′ October 23 transit STR 1311

Location·Markab is at the southwest corner of the Great Square of Pegasus and it is slightly brighter than Scheat, which lies 13 degrees to the north.

Meaning·Markab appears to have first been used in reference to this star during the sixteenth century. The name may have been borrowed from Markeb, a fourth-magnitude star in the constellation Puppis.

Lore·In ancient Greece, this star helped to mark the back of the Flying Horse, Pegasus. The figure of the horse was portrayed upside down, with the Great Square representing his body.

Description•Markab is a giant blue-white star with a magnitude of 2.5. Its surface temperature is about 10,000°K, and it has a luminosity 90 times greater than the sun's.

The parallax of Markab is 0.030 arcsecond, which indicates a distance of 33 parsecs, or about 100 light-years. This distance is close to the limit at which the method of trigonometric parallax is reliable.

Nearby Feature

THE GREAT SQUARE OF PEGASUS

The Great Square is one of the best-known asterisms due to its distinct shape and prominent position high in the southern sky during autumn evenings.

Once you have learned to identify the four stars that form this Square (Markab, Scheat, Alpheratz, and Algenib), you may enjoy testing your vision and the clarity of the sky by counting stars within this asterism. Upsilon Pegasi, with a magnitude of 4.4, is located in the northwest quadrant of the Square, 1 degree distant from Tau Pegasi, a star of magnitude 4.6. These are the two brightest stars within the Great Square, and if they are the only stars you can see within the asterism, 4.6 is your limiting magnitude at that particular time. If you are able to count five stars inside the Great Square of Pegasus, the limiting magnitude is about 5.0, and if you count thirteen, the limit is about magnitude 6.0.

The usual naked-eye limit is about magnitude 6.5; one would have to see about thirty-five stars inside the Square in order to reach this "limit." Julius Schmidt, a director of the Athens Observatory during the nineteenth century, is reported to have observed 102 stars within the Great Square, which corresponds to a magnitude limit of nearly 7.4!

See the constellation map for Pegasus (page 238) and Andromeda (page 270).

Aquarius, *the Water Carrier*

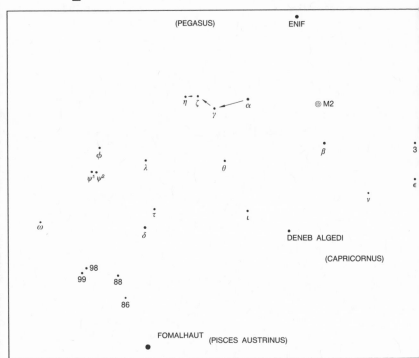

ack-KWAIR-ee-us
Aqr
Aquarii

Aquarius is a faint zodiac constellation whose principal feature is an asterism known as the Water Jar and which contains mostly fourth-magnitude stars.

Season•September – January

Sky Track•Aquarius rises in the east, crosses the meridian about halfway between the horizon and the zenith, and sets in the west.

Lore•The stars of Aquarius have been known since Babylonian times to be representative of a man pouring water from an urn. In ancient Mesopotamia the sun was seen in the same portion of the sky as this constellation during the rainy winter months, whence came the association with water. Aquarius is in a part of the sky known to Mesopotamians as the Sea, and to the early Greeks as the Water. Most of the surrounding constellations also have associations with water because of the weather in the ancient Middle East when the sun was in this section of the sky.

A sprinkling of more than 30 stars with magnitudes of four and fainter represents the water flowing from the Water Jar in Aquarius.

Description•Binoculars are a help in finding the Water Jar and the rest of this constellation. Sweep across the sky with your binoculars beginning at the star Enif in Pegasus, in the direction of Fomalhaut in Piscis Austrinus. The Water Jar is located about 15 degrees south of Enif on this path. The faint stars pouring from the Jar lie to the southeast of the Water Jar and east of the imaginary line from Enif to Fomalhaut.

Nunki   August 21

South-Southeast   South   South-Southwest

Nunki	2.1 magnitude	−2.5 absolute
NUN-kee	blue-white B2	magnitude
Sigma, σ	main sequence V	18h 55m −26° 18′
Sagittarii, page 248	250 light-years	August 21 transit
	+0.013 −0.054 −11	STR 1315

Location•Second-magnitude Nunki lies at the top of the handle of the Teapot in Sagittarius.

Meaning•The name Nunki is believed to have originated in Mesopotamia as part of a description meaning the Star of the Proclamation of the Sea. This "sea" was a portion of the sky occupied by the water-associated constellations Aquarius, Capricorn, Delphinus, Pisces, and Piscis Austrinus.

Lore•The region of sky whose arrival Nunki heralds was described by the Greek poet Aratus as the Water. In the traditional portrayal of Sagittarius, Nunki marks the arrow's vane being held by the Archer as he draws back his bowstring.

Nunki is the brightest star in the asterism of the Milk Dipper.

Description·Nunki is a blue-white main-sequence star with an apparent magnitude of 2.1. Its surface temperature is estimated to be 23,000°K, and its luminosity is equal to the brightness of 850 suns. Stars of this spectral type usually have absolute magnitudes of about −2.5, and comparison of this figure with Nunki's apparent magnitude provides an estimated spectroscopic parallax of about 250 light-years.

Main-sequence stars of Nunki's spectral type are estimated to have masses about 4.5 times greater than that of the sun. Such stars are believed to have main-sequence life expectancies in the order of about 50 million years. This means that Nunki probably became a star sometime after the demise of the dinosaurs on earth.

Ascella   August 23

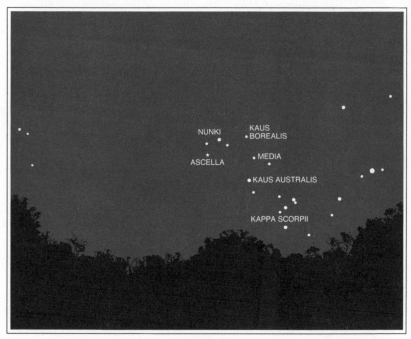

South-Southeast	South	South-Southwest

| Ascella
ah-SELL-ah
Zeta, ζ
Sagittarii, page 248 | 2.6/3.2 magnitude
blue-white A2
giant III
130 light-years
−0.014 −0.001 +22 | +0.2 absolute
magnitude
19h 03m −29° 53′
August 23 transit
STR 1323 |

Location•Ascella is the last of our featured stars in Sagittarius to cross the meridian. It marks the lower juncture of the Teapot with its handle, and it is a third-magnitude star.

Meaning•The name Ascella is derived from a Latin description of the Arm of the Centaur.

History•Ascella consists of the blended light of two close double stars whose binary nature was discovered by W. C. Winlock of the United States Naval Observatory in 1867.

The Naval Observatory has long served a mandate to provide information useful in the areas of navigation and time determination. Simon Newcomb, one of its early directors, refined theoretical standards used to measure time to a high degree of accuracy through astronomical observations.

Description·The two components of Ascella are nearly equal in brightness. They are a blue-white giant of magnitude 3.2 and a blue-white subgiant star of magnitude 3.4. The pair has an orbital period of just over 21 years. Their current separation is about 0.4 arcsecond.

Spectral class A2 stars feature strong lines of hydrogen. Absorption lines of neutral helium and ionized metals are also characteristic of these stars. The hydrogen lines in the visual part of the spectrum, and also lines of ionized titanium and iron, are used to estimate luminosity differences in class A2 stars. Ascella, for example, shows lines of intermediate width between the narrow lines seen in A2 supergiants such as Deneb and broad luminosity criteria lines characteristic of A2 supergiants such as Lambda Ursae Majoris.

The proper motion of Ascella is westward 0.014 arcsecond and southward 0.001 arcsecond per year. This star moves away from the sun at the rate of 22 kilometers per second. The distance of the Ascella system from earth is about 130 light-years.

Ruchbah   August 25

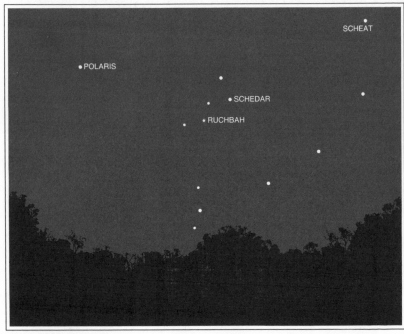

North-Northeast   Northeast   East-Northeast

Ruchbah	2.7 magnitude	+0.5 absolute
RUCK-bar	blue-white A5	magnitude
Delta, δ	giant-subgiant III–	01h 26m +60° 14'
Cassiopeiae, page	IV	November 28 transit
242	90 light-years	STR 1334
	+0.300 −0.045 +7	

Location•Ruchbah is at the lower left point of the W-shaped asterism in Cassiopeia. This third-magnitude star follows in the train of Caph, Schedar, and Gamma Cassiopeiae as the constellation moves counterclockwise around the North Celestial Pole. This apparent motion is a reflex of the earth's rotation and orbital motion.

Meaning•The name of this star is adapted from its place in the ancient constellation figure, at the Knee of the Woman in the Chair.

History•In 1669, French astronomer Jean Picard used Ruchbah as one of the reference stars in his survey of the size and shape of the earth. In a refinement of the method used by Eratosthenes, Picard measured the distance along a north-south line required to change the apparent altitude of Ruchbah by 1 degree of arc. This project is said to have made the first use of a telescope in geodesic research.

Description•Ruchbah is classified as a blue-white giant or subgiant star. Its trigonometric parallax indicates a distance of 90 light-years and a luminosity 50 times greater than that of the sun. The star's spectrum indicates a surface temperature of about 8,500°K.

The apparent magnitude of Ruchbah has a usual value of 2.68 and a variable range between magnitudes 2.68 and 2.76, which is indicative of a partial eclipse by a companion star. The brightness of Ruchbah varies over the period of 759 days.

The proper motion and radial velocity of Ruchbah are very similar to those of stars in the Hyades open cluster in the constellation Taurus. It has therefore been suggested that Ruchbah is an outlying member of this star cluster.

Nearby Feature

H AND CHI PERSEI: A DOUBLE STAR CLUSTER:
H (02h 18m + 57° 04′), CHI (02h 21m + 57° 02′)

Twelve degrees from Ruchbah, along a line extended from this star and Gamma Cassiopeiae, is the lovely Double Cluster in the constellation Perseus. The Double Cluster was described by both Hipparchus and Ptolemy. It lies in a beautiful section of the Milky Way, and the view of this region through binoculars is spectacular.

Visible to the naked eye on a clear night as a hazy spot of light, these two clusters, each containing hundreds of stars, are both at the center of a huge region of giant stars with a diameter of several hundred light-years at a distance from earth of about 7,500 light-years.

The brightest stars in clusters h and Chi Persei are blue-white supergiants of spectral types A and B, with apparent magnitudes of about 6.5. If it were not for the absorption of light by dust on its journey to earth, stars in the Double Cluster would appear about 1.6 magnitudes brighter to us.

Presence of these A- and B-type supergiants, which are prodigious in their use of fuel, indicates that the Double Cluster is quite young, relative to the age scale of stars. It has been suggested that h and Chi formed in a cloud of Milky Way dust and gas about 6 and 12 million years ago, respectively.

See the constellation maps for Cassiopeia (page 242) and Perseus (page 292).

Alpheratz August 26

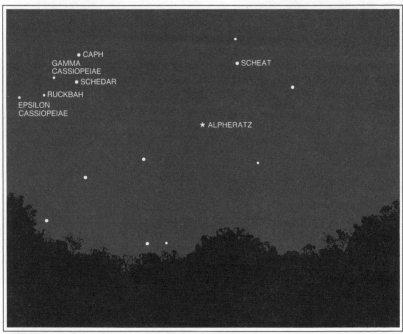

Northeast	East-Northeast	East
Alpheratz al-FEE-rats Alpha, α Andromedae, page 270	2.0 magnitude blue-white B9 peculiar 130 light-years +0.137 −0.158 −12	−0.9 absolute magnitude 00h 08m +29° 05′ November 8 transit STR 1337

Location•Second-magnitude Alpheratz is formally a part of the constellation Andromeda, although it serves to mark the northeast corner of the Great Square of Pegasus.

Meaning•Alpheratz appears to have first been used as a proper name for this star by Johann Bayer in 1603, having been adapted from a medieval description of Beta Pegasi as the Shoulder of the Horse.

Lore•In classical sky mythology this star represented the Head of the Woman in Chains, Andromeda. The star has traditionally been associated with both Andromeda and Pegasus; however, in 1928, the International Astronomical Union defined formal constellation boundaries and assigned Alpheratz to Andromeda.

In England, Alpheratz is popularly known as Andromeda's Head.

The stars Caph in Cassiopeia, Alpheratz, and Algenib in Pegasus are sometimes known as the Three Guides since they lie close to the great circle that marks zero hours of right ascension in this part of the sky. This line, known as the equinoctial colure, passes between the celestial poles through the point of the vernal equinox, where the ecliptic intersects the celestial equator.

Description·Alpheratz is a blue-white star of average magnitude 2.0. Its spectrum is classified as peculiar and shows unusually strong lines of manganese and gallium. It is considered to be a magnetic-spectrum variable with a magnitude range from 2.02 to 2.06 and a period of 0.964 day. Cor Caroli in Canes Venatici and Mirzam in Canes Major are other examples of this type of variable.

The distance to Alpheratz is estimated to be about 130 light-years, which would make its average luminosity 200 times that of the sun.

This star is a spectroscopic binary whose period is 96.7 days.

Andromeda, *the Chained Lady*

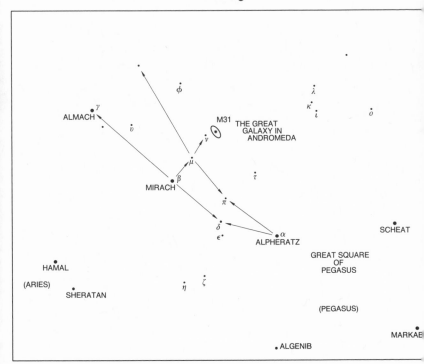

BRIGHT STARS OF ANDROMEDA

August 26	Alpheratz	2.0	blue-white	peculiar
September 7	Mirach	2.1	yellow-orange	giant
September 17	Almach	2.2	yellow	giant

an-DROM-eh-da
And
Andromedae

Evening Season•September – February

Sky Track•The constellation of Andromeda rises in the northeastern sky, passes overhead, and then moves towards the northwest.

Lore•In Greek mythology Andromeda was the Princess of Ethiopia, daughter of King Cepheus and Queen Cassiopeia. Cassiopeia had boasted that she was more beautiful than the lovely Nereids, daughters of Nereus, god of the Mediterranean Sea. In an act of divine retribution, a sea serpent was sent to plague the coast of Ethiopia. After numerous inhabitants had been killed by this creature, an oracle suggested that only the sacrifice of Princess Andromeda would bring relief from the monster's ravages. After much hesitation, Cepheus and Cassiopeia were forced to give up their daughter, and Andromeda was chained to a rocky promontory at the edge of the sea to await her fate.

At this point, the mighty hero Perseus appeared, on his way home with the head of the Medusa. Perseus discovered what was happening and stayed with Andromeda until the monster arrived. With one stroke of the sword given to him by Hermes, Perseus decapitated the serpent and its body slid under the waves. Andromeda was released and went with Perseus to become his bride.

Description·Andromeda's head is represented by the star Alpheratz, which also marks the northeastern corner of the Great Square of Pegasus.

Other stars in the constellation Andromeda extend to the northeast from Alpheratz in two diverging lines. The bright stars Mirach and Almach are in the more southerly line, which points towards Mirfak, the brightest star in the constellation Perseus. Fourth- and fifth-magnitude stars form the northern chain of stars in Andromeda and can be observed under clear conditions.

The Great Galaxy in Andromeda, M31, is the major feature of this constellation. At a distance of about 2.1 million light-years from earth, it is both the nearest large spiral galaxy to our Milky Way and also the most distant object generally visible with the naked eye. It appears as a hazy oval about 1 degree in length to observers without optical aid.

The Andromeda galaxy is located about 10 degrees of arc to the northwest of Mirach. (See "Nearby Feature" of Mirach, September 7, for more on this galaxy.)

The second-magnitude star Almach is a lovely double whose components have contrasting colors of yellow and blue-white, with a separation of about 10 arcseconds. It is possible to observe the components of Almach with a small telescope.

Deneb Algedi   September 3

East-Southeast	Southeast	South-Southeast
Deneb Algedi DEN-ebb al-JEE- dee Delta, δ Capricorni, page 274	2.9 magnitude blue-white A7 subgiant IV 37 light-years +0.262 −0.294 −6	+2.6 absolute magnitude 21h 47m −16° 08′ October 3 transit STR 1368

Location• Deneb Algedi is a third-magnitude star located 28 degrees of arc directly to the south of Enif. It is the brightest star in Capricornus and it lies at the eastern corner of a large triangular asterism formed by stars in this constellation.

Meaning•The name Deneb Algedi comes from a description, found in medieval astrolabe tables, that meant the Tail of Capricornus. This corresponds to this star's location in traditional portrayals of the Sea Goat.

History•The planet Neptune was discovered about 4 degrees to the northeast of Deneb Algedi by the German astronomer Johann G. Galle on the night of September 23, 1846. He was searching in a region of sky suggested by the French astronomer Urbain Le Verrier, who, along with John C. Adams of England, had predicted its location through mathematical analysis of orbital characteristics of the planet Uranus.

Description•Deneb Algedi is a blue-white subgiant star with an average apparent magnitude of 2.9. Its arrival in the evening sky around Labor Day coincides with the unofficial conclusion of summer and the advent of cooler weather. At this time of year, the cool, clear nights are well suited for stargazing.

The trigonometric parallax of Deneb Algedi is 0.087 arcseconds, which corresponds to a distance of 11.5 parsecs, or about 37 light-years. A knowledge of its distance permits the calculation of this star's average absolute magnitude at approximately 2.6. This is equivalent to a luminosity about 8 times greater than that of the sun. The spectrum of Deneb Algedi indicates a surface temperature of 8,000°K.

Deneb Algedi is an eclipsing binary system with a slight brightness variation between magnitudes 2.82 and 3.05 over a period of 1.023 days. The components have individual magnitudes of 3.2 and 5.2, with a separation calculated to be about 0.0018 arcsecond.

Nearby Feature

ALGIEDI: A BINOCULAR DOUBLE (20h 18m − 12° 31′)

Twenty-two degrees to the west of Deneb Algedi lies a wide optical double whose components have magnitudes of 3.6 and 4.0 with a separation of 378 arcseconds, equal to one-fifth the apparent diameter of the full moon. The gap between these two yellow-white stars is easily visible with binoculars. Algiedi, also known as Alpha Capricorni, lies at the northwest corner of the triangle-shaped asterism of Capricornus.

The brighter star in this pair is designated as Alpha[2] and is 95 light-years from earth. The distance of Alpha[1] is estimated to be about 2,000 light-years from us.

Capricornus, *the Sea Goat*

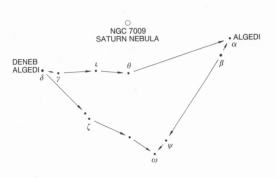

BRIGHT STAR OF CAPRICORNUS

September 3	Deneb Algedi	2.9	blue-white	subgiant

CAP-rih-CORE-nus
Cap
Capricorni

Evening Season•September – October

Sky Track•Capricornus rises into view in the southeast and remains near the southern horizon before setting towards the southwest.

Lore•This constellation has most often been described in Western sky mythology as a sea goat or simply as a goat. When portrayed as a sea goat, Capricornus is usually shown with the head and body of a goat but with the tail of a fish rather than hind legs.

Description•When seen through clear, dark skies, Capricornus presents an elegantly symmetrical figure of faint fourth- and fifth-magnitude stars.

Alpha Capricorni, also known as Algedi, is a binocular double located at the northwest corner of the constellation figure.

Nearby Feature

THE SATURN NEBULA (21h 03m −11° 28′)

This ninth-magnitude planetary nebula has an NGC (New General Catalogue) number of 7009 and appears as a bright green oval when seen through a 6-inch or larger telescope. The object's widest apparent diameter is about 45 arcseconds. William Parsons gave it the name Saturn Nebula because, with his 72-inch telescope, two rays were seen extending from its center in a way that made this object resemble the planet Saturn.

The Saturn Nebula is located in the southern part of the constellation Aquarius, east of Deneb Algedi, at a distance of about 3,000 light-years from earth.

Algenib   September 6

East-Northeast   **East**   **East-Southeast**

Algenib	2.8 magnitude	−3.7 absolute
al-JEE-nib	blue-white B2	magnitude
Gamma, γ	subgiant IV	00h 13m +15° 11′
Pegasi, page 238	650 light-years	November 9 transit
	+0.003 −0.007 +4	STR 1379

Location•Algenib marks the southeast corner of the Great Square of Pegasus and it is the last star in this asterism to rise into sight. It is a third-magnitude object located 14 degrees of arc to the south of Alpheratz.

Meaning•The name Algenib appears to have had no ancient or medieval association with Pegasus. Due to the similarity of the names Pegasus and Perseus, it is possible that Algenib was erroneously taken from an old name for Alpha Persei.

Lore•Star maps of the past generally showed Algenib near a wing tip of Pegasus.

Along with the stars Caph in Cassiopeia and Alpheratz in Andromeda, Algenib is one of the Three Guides that help to mark the location of the zero-hour line of right ascension.

History•During 1911, astronomer K. Burns of the Lick Observatory discovered that the radial velocity of Algenib varies periodically, which indicates that the star's surface is pulsating slightly. By 1952 a period of about 3 hours and 38 minutes had been measured.

Description•Algenib is a blue-white subgiant variable of the Beta Cephei type with an average apparent magnitude of 2.8. Such stars are believed to have recently begun their evolution away from the main sequence and are increasing in diameter as additional energy is produced through the fusion of helium in their cores.

The surface temperature of Algenib is estimated to be 23,000°K, and its distance from earth is calculated to be about 650 light-years.

Mirach   September 7

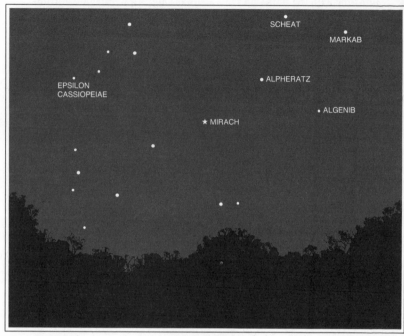

Northeast   **East-Northeast**   **East**

Mirach	2.1 magnitude	+0.1 absolute
MY-rack	yellow-orange M0	magnitude
Beta, β	giant III	01h 10m +35° 37′
Andromedae, page	75 light-years	November 24 transit
270	+0.179 −0.109 +3	STR 1382

Location•Second-magnitude Mirach may be found 14 degrees to the northeast of Alpheratz in the constellation of Andromeda.

Meaning•The name Mirach possibly evolved from an Arabic term that described the Apron of Andromeda.

History•E. E. Barnard discovered a fourteenth-magnitude companion to Mirach at a distance of 28 arcseconds while observing from Yerkes Observatory in 1898. The two stars have the same proper motion and are probably bound together by gravity.

Description•Mirach is a yellow-orange giant star with an apparent magnitude of 2.1. Measurement of this star's trigonometric parallax indicates a distance from earth of 75 light-years. Mirach's surface temperature is a comparatively low 3,200°K, and the star's spectrum features absorption lines of neutral metals. With an absolute magnitude estimated to be +0.1, Mirach is 75 times more luminous than the sun.

Two twelfth-magnitude optical companions lie about 80 arcseconds from Mirach.

Nearby Feature

M31: THE GREAT GALAXY IN ANDROMEDA (00h 42m +41° 09')

This is the most distant object that can be seen clearly with the naked eye. It is a vast spiral galaxy, similar to the Milky Way, and lies at a distance of about 2 million light-years. The Andromeda galaxy appears as a hazy oval centered about 9 degrees to the northwest of Mirach, and seen with binoculars, it extends several degrees.

This galaxy was described as the Little Cloud by the tenth-century Persian astronomical writer Al Sufi. The German astronomer Simon Marius first described its telescopic appearance after having observed the galaxy in December of 1612.

When observed through small telescopes, M31 looks like a softly glowing cloud, with no hint of resolution into stars. Until this century the object was known as the Great Nebula in Andromeda, and there were some suggestions that it was a solar system in the process of formation.

In 1885, a new star appeared in the heart of M31 and reached an apparent magnitude of 7.2 before it began to fade. At that time, it was assumed that this was an ordinary nova, similar to those that had been observed from time to time in the Milky Way. The brightness of this star, designated S Andromedae, indicated that it was probably only a few thousand light-years from earth, and this was taken as evidence that M31 was some sort of nebula in the known realm of the universe. After the true distance of M31 had been determined, it was evident that S Andromedae must have been a brilliant supernova, thousands of times brighter than an ordinary nova.

In 1923, Edwin Hubble discovered a number of variable stars in M31 and, by their characteristic changes in brightness, identified them as Cepheid variables. From the research on these stars performed by Henrietta Leavitt in 1912, it was known that Cepheids are giant stars whose periods of variation are directly related to their luminosities, yet the specimens seen in M31 had apparent magnitudes of only about +17, which indicated distances far in excess of the known diameter of the Milky Way star system. As a result, the extragalactic location and immense size of M31 were generally accepted, and the object came to be known as the Great Galaxy in Andromeda rather than as the Great Nebula. With this recognition, the acknowledged volume of the observed universe increased many thousands of times.

Almach   September 17

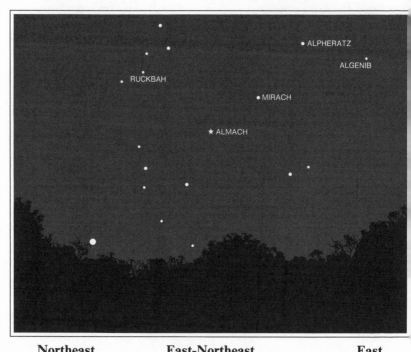

Northeast	East-Northeast	East

Almach	2.2/2.3 magnitude	−2.1 absolute
al-MAK	yellow K3	magnitude
Gamma, γ	giant II	02h 04m +42° 20′
Andromedae, page	250 light-years	December 7 transit
270	+0.046 −0.048 −12	STR 1422

Location•Almach is the third star along a line that begins with Alpheratz and passes through Mirach in Andromeda. These are second-magnitude stars, of nearly equal brightness.

Meaning•The name Almach is believed to have been taken from an Arabian description of this star as the Desert Lynx.

Lore•In classical sky lore, Almach marked the left foot of the Chained Lady, Andromeda.

History•Almach was found to be a lovely double star by the German astronomer Johann T. Mayer in the middle of the eighteenth century.

In October of 1842, Wilhelm von Struve, first director of the Imperial Russian Observatory at Pulkowa, discovered that Almach's fifth-magnitude companion actually consists of two stars with a separation of less than 1 arcsecond.

Description•Almach is a yellow giant star whose apparent magnitude is 2.3. Its distance is estimated to be about 250 light-years and its surface temperature is approximately 4,000°K. Stars of this type are about 650 times more luminous than the sun.

The companion lies at a distance of 9.6 arcseconds and consists of two blue-white main-sequence stars, of magnitudes 5.5 and 6.3. The maximum separation of these close neighbors is about 0.55 arcseconds, which was attained in 1982. Almach and its companions appear to the naked eye as a single star of apparent magnitude 2.2.

It has been found that the brighter of these very close stars is itself a spectroscopic double, so that the Almach system contains a total of four stars.

Almach is considered one of the most beautiful double stars. The yellow and blue-white color contrast and the separation are fine sights in telescopes of all sizes.

The autumn season begins on or about September 23, when the sun appears to move south across the celestial equator at a right ascension of 6 hours in the constellation of Virgo.

Sheratan   September 28

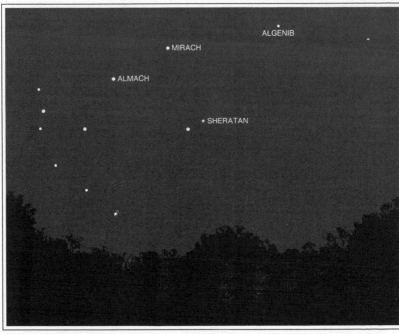

East-Northeast	East	East-Southeast
Sheratan SHARE-ah-tan Beta, β Arietis, page 284	2.6 magnitude blue-white A5 main sequence V 44 light-years +0.097 −0.108 −2	+2.0 absolute magnitude 01h 55m +20° 48′ December 20 transit STR 025

Location•Third-magnitude Sheratan is located 16 degrees of arc to the southeast of Mirach.

Meaning•The name Sheratan was taken from an early Arabian description of this star and its neighbor Mesarthim as the Two Signs.

Lore•Sheratan and Mesarthim were known in ancient India as the Two Horsemen and in Persia as the Protecting Pair. The name Mesarthim comes from a Hebrew description of stars in Aries.

History•Sheratan and Mesarthim marked the site of the vernal equinox about 2,000 years ago. The arrival of the sun at this point of the ecliptic indicated the start of a new year during ancient times in the classical world. As a result, these stars in Aries were important celestial signs, and even today the constellation is considered to be the first in the order of the zodiac, even though precession has carried the vernal equinox about 30 degrees of arc to the west, into the constellation Pisces.

Hermann K. Vogel found from observations he made in 1903 that Sheratan is a spectroscopic binary.

Description•Sheratan is a main-sequence star. Its spectrum indicates a surface temperature of 8,500°K for this blue-white star. Its apparent magnitude is 2.6. Sheratan's trigonometric parallax of 0.074 arcseconds indicates a distance of 13.51 parsecs, or about 44 light-years. Knowledge of Sheratan's distance permits us to calculate this star's luminosity to be about 13 times that of the sun.

Main-sequence stars with this luminosity are estimated to have about twice the mass of the sun. Such stars are believed to have main-sequence life expectancies on the order of 1.5 billion years. A star leaves the main sequence and swells to become a giant when the helium produced in its core reaches about 12 percent of the star's total mass. In the giant stage, helium and other elements heavier than hydrogen begin to release energy through various nuclear reactions.

Aries, *the Ram*

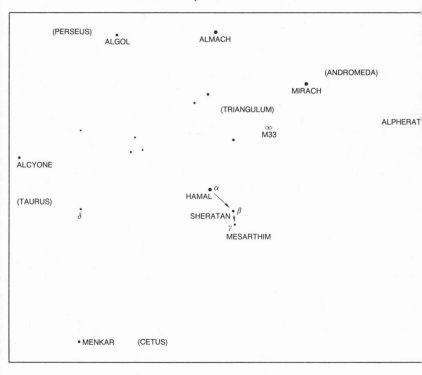

BRIGHT STARS OF ARIES

September 28	Sheratan	2.6	blue-white	main sequence
September 29	Hamal	2.0	yellow	giant

AIR-ease
Ari
Arietis

Evening Season•October – March

Sky Track•This constellation is seen to move across the sky from east to west, crossing the meridian halfway between the horizon and the zenith.

Lore•Aries heads the list of zodiac constellations since it marked the site of the vernal equinox during the period between approximately 1800 B.C. and A.D. 1. In ancient times this point on the ecliptic indicated the sun's location in the sky at the start of both the spring season and the new year. This point continues to be called the First Point of Aries, even though the precession of the earth's axis has carried the actual location of this point into the constellation Pisces.

The historian Josephus wrote that the sun was situated in Aries at the time of the year when the people of Israel were released from Egyptian bondage. An ancient belief in Mesopotamia held that the world was created while the sun was located near these stars. In Assyria, the constellation represented an altar and a sacrificial ram.

In classical Greek mythology, Aries is associated with the Ram of the Golden Fleece. It was said that this ram, with fleece of the purest gold, had been sent by Hermes to rescue a prince named Phrixus whose life was threatened by palace intrigue. The ram carried Phrixus to safety by flying to the land of Colchis on the Black Sea. Phrixus then sacrificed the ram and gave its Golden Fleece to King Aetes, in thanks to him for having provided sanctuary in Colchis.

Jason, a young prince whose father's kingdom had been usurped by King Pelias, was later challenged by this usurper to capture the Golden Fleece. Jason organized the argonautic expedition, whose members included the greatest heroes in all of Greece: Hercules, Orpheus, Castor, and Pollux. After many adventures, the Argonauts reached Colchis and, with the essential help of Princess Medea, took the Golden Fleece from Aetes.

Description·Although Aries is a small constellation, the proximity of Hamal and Sheratan presents a distinctive pair of stars that enables us to find the Ram with relative ease. It is located about 20 degrees to the southeast of the bright stars in Andromeda.

The Ram's horns are often shown represented by the bent line of the stars Hamal, Sheratan, and fifth-magnitude Mesarthim.

Pisces, *the Fish*

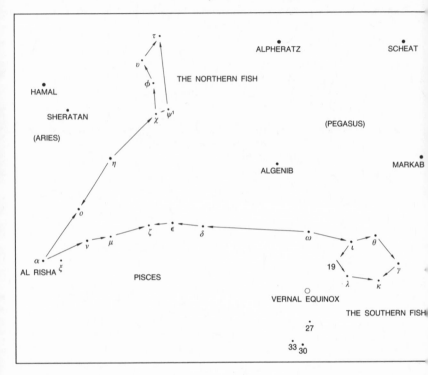

PIE-sees
Pic
Piscium

Pisces is a rather large but faint constellation in the zodiac whose brightest star is only of the fourth magnitude. It is located to the south and east of Pegasus.

Season•October – January

Sky Track•The faint stars of Pisces are spread across the sky near the celestial equator. As a result, this constellation rises near the eastern point of the horizon and sets at the western point. When it passes the meridian it is about halfway between the southern horizon and the zenith.

Lore•These stars have been known since Mesopotamian times as the Fish. The constellation is shown as a pair of fish, one south of the Great Square of Pegasus called the Southern Fish, and another to the east of the Square known as the Northern Fish. These celestial fish are shown bound together by cords knotted at the star Al Risha, Alpha Piscium.

The Southern Fish is represented by a circlet of fourth- and fifth-magnitude stars, while the Northern Fish is also a faint asterism, which is roughly egg-shaped.

In Greek sky lore, Aphrodite and her son Eros were bathing by a river bank when they became frightened by the approach of the terrible monster, Typhon. In order to escape, they transformed themselves into a pair of fish bound together so that they would not become separated. In memory of this event, the images of these two fish were placed in the sky.

Description • Most of the stars in Pisces have magnitudes of only four or five and excellent seeing conditions are needed for their observation with the naked eye. With binoculars, however, the constellation may be traced under less than perfect conditions.

Around March 21 each year, the sun is located in Pisces at a point on the celestial sphere known as the vernal equinox. The arrival of the sun at this point signals the start of the spring season in the northern hemisphere.

The vernal equinox lies at the intersection of the great circles of the celestial equator and the zero-hour colure line of right ascension. This spot is sometimes called the First Point of Aries, even though it has been two thousand years since precession carried the vernal equinox out of the constellation Aries and into Pisces.

Hamal   September 29

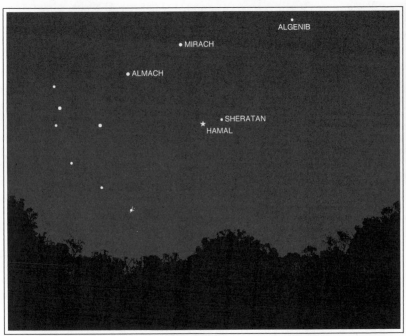

East-Northeast   **East**   **East-Southeast**

Hamal	2.0 magnitude	+0.2 absolute
HAM-al	yellow K2	magnitude
Alpha, α	giant III	02h 07m +23° 28′
Arietis, page 284	75 light-years	December 8 transit
	+0.190 − 0.144 − 14	STR 031

Location•Hamal, a second-magnitude object, is the brightest star in the constellation Aries and lies 4 degrees northeast of Sheratan.

Meaning•The name Hamal was adapted from an Arabian name for the constellation Aries.

Lore•According to some sources, Hamal was known in Mesopotamia as the Horn Star and in Babylon as the Messenger of Light.

Description•Hamal's apparent magnitude is 2.0 and its trigonometric parallax indicates a distance of 75 light-years. This giant star is yellow and has a surface temperature of about 4,200°K. Calculation of Hamal's absolute magnitude, +0.2, indicates that this star is about 70 times more luminous than the sun.

Nearby Feature

M33: SPIRAL GALAXY (01h 33m + 30° 33′)

M33 in the constellation Triangulum is located 10 degrees northwest of Hamal. A few skilled observers claim the ability to see this galaxy when viewing conditions are excellent, and for them it is the most distant object visible with the naked eye. A small telescope will show M33 when the sky is clear and free from most light pollution, but it is quite pale.

This spiral galaxy in Triangulum is a member of the Local Group of Galaxies, along with the Milky Way, the Great Galaxy in Andromeda, and about seventeen other smaller galaxies. Its distance is estimated to be about 2.3 million light-years from us.

See the Aries constellation map, page 284.

When looking for objects at the threshold of visibility, it is helpful to use the method of "averted vision," whereby you stare at a point 10 or 20 degrees to one side of your intended object. When you do this, the object's faint light falls on a more sensitive part of your retina than if you had stared directly at the object.

Mirfak

October 2

| North-Northeast | Northeast | East-Northeast |

Mirfak	1.8 magnitude	−4.3 absolute
MERE-fak	white F5	magnitude
Alpha, α	supergiant Ib	03h 24m +49° 52m
Persei, page 292	500 light-years	December 28 transit
	+0.025 −0.022 −2	STR 0041

Location•Second-magnitude Mirfak is the brightest star in the constellation Perseus, the Hero. It may be found 17 degrees beyond Almach, in a line of bright stars that also includes Mirach and Alpheratz in Andromeda.

Meaning•The name Mirfak originated with an early Arabian description of stars in Perseus as the Elbow of the Pleiades and was adapted from the word for elbow.

Description•The white supergiant star Mirfak lies at a distance of about 500 light-years from earth. It is 4,400 times more luminous than the sun and has a surface temperature of approximately 6,500°K. With an apparent magnitude of 1.8, Mirfak lies along the delicate tracery of star chains seen in the constellation Perseus.

Nearby Feature

ALPHA PERSEI STAR ASSOCIATION

Mirfak is found in a splendid little patch of sky rich in blue-white stars belonging to the Alpha Persei Association. Observers using binoculars will see this highly attractive group, which is centered just to the south of Mirfak. Proper-motion studies have identified over 100 members of this star association, which features young, very luminous, and massive type-B stars.

Faint luminosity surrounding these stars is a remnant of the cloud of dust and gas from which members of this cluster formed. Although most of this natal nebula has disappeared, not enough time has passed for bright cluster members to have burned themselves out or to have moved apart from each other. This is why such an association of spectral types O and B, blue-white, stars is so attractive.

Perseus, *the Hero*

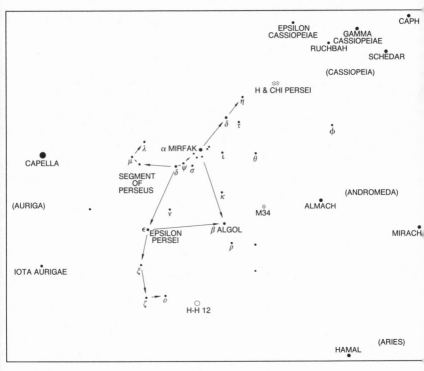

BRIGHT STARS OF PERSEUS

October 2	Mirfak	1.8	white	supergiant
October 4	Algol	2.2	blue-white	main sequence
October 17	Epsilon Persei	2.9	blue-white	main sequence

PURR-see-us
Per
Persei

Evening Season•October – March

Sky Track•Perseus follows an apparent path across the sky from the northeast to northwest. It is at the zenith as it crosses the meridian when seen from midnorthern latitudes.

Lore•Perseus, the Hero, is one of a group of constellations, seen on autumn evenings, called the Royal Family of the Sky. This group includes Perseus, Andromeda, Cepheus, and Cassiopeia, whose legends are also entwined with those of Pegasus and Cetus.

Perseus was said to have been born from the union of Zeus and a princess of Argos named Danae. His grandfather, King Acrisius, had once been told by the Delphic Oracle that his daughter's son would kill him. As a precaution, the king set Danae and Perseus adrift in a wooden chest, with the hope that the sea would carry them away forever.

However, the next day the chest was blown ashore on an island, where it was found by a kind fisherman named Dictys. He and his wife took Danae and Perseus into their home, where they lived happily for many years.

As Perseus grew into manhood, learning the skills of a fisherman, his still-beautiful mother became a source of attraction for the island's king, Polydectes, an evil brother of Dictys, the fisherman. Polydectes wanted to have Danae as his wife but saw Perseus as a source of distraction and trouble. In order to be rid of Perseus, the king let it be known that, above all else, he desired the head of the extremely dangerous Medusa as a wedding gift. Proud Perseus had nothing else to offer and bravely accepted this nearly impossible challenge from Polydectes. The Medusa was one of three hideous creatures known as Gorgons, who had body scales and wings, as well as a tangle of snakes for hair. To make matters worse, anyone who looked at them would instantly be transformed into stone.

Perseus took the first ship to Greece, where he was comforted with news provided by the talking oaks of Dodona that he was under the protection of the gods. Hermes then came to Perseus and presented him with a magic sword, which could be neither broken nor bent. The goddess Athena appeared and gave Perseus her polished bronze shield, which he would use to view his prey indirectly.

Hermes traveled with Perseus to the Northland where they tricked three women known as the Gray Sisters into telling them where to find the nymphs of the North. The Northern nymphs provided a great feast for Perseus and gave him a cap with the power of invisibility, a pair of winged sandals, and a leather sack.

Fully prepared, Perseus flew to the distant land of the Gorgons and, harmlessly viewing their reflections in Athena's mirrorlike shield, found them asleep. Athena and Hermes were at his side and they pointed out Medusa, the only mortal Gorgon. Guided by the reflection, Perseus descended, cut off the head, and placed it in his sack. The surviving Gorgons awoke, but Perseus put on the cap of invisibility and escaped. As Perseus departed, Neptune created the flying horse, Pegasus, from a mixture of Medusa's blood and sea foam.

Perseus flew along the Ethiopian coast and was able to rescue Princess Andromeda after killing the serpent that had threatened her. King Cepheus and Queen Cassiopeia happily gave their blessings, and Andromeda sailed with Perseus to his home island as his bride.

Upon his return, Perseus learned that his mother had rejected Polydectes' marriage proposal and the king had forced Danae and Dictys into hiding. Perseus tracked down the king and his minions in the middle of a great palace banquet. Standing at the entrance, wearing Athena's glorious armor, Perseus caught everyone's eye. He then withdrew the Medusa's head from its sack, turning everyone into stone.

After the islanders gratefully accepted Perseus' recommendation and made Dictys their new king, Perseus, Andromeda, and Danae returned to Greece in hopes of being reconciled with Acrisius, who had set the young mother and child adrift so many years before. However, they learned that Acrisius had been sent into exile and no one knew where he was.

Having heard of a great athletic festival, Perseus journeyed north to take part in these games. While he was throwing a discus, the projectile landed in the crowd and killed one of the spectators. It turned out that the dead man was Perseus' exiled grandfather, who had been visiting the local king, and with his death the Delphic prediction had finally come true.

Perseus and Andromeda went on to live a happy life together, and the myths tell us they were the great-grandparents of the mighty hero Hercules.

Description•The constellation Perseus is found in a section of the Milky Way that is very lovely when seen through binoculars.

An asterism known as the Segment of Perseus consists of a J-shaped chain of stars, including Mirfak, which extends towards the northeast. Another pattern, a reversed J, stretches to the south from Mirfak and includes the star Epsilon Persei.

This constellation contains several interesting features, including the double-star cluster h and Chi Persei, located near the W of Cassiopeia; and the eclipsing binary star Algol.

All of the bright stars seen in our sky belong to the local Orion arm of the Milky Way Galaxy, the same spiral arm that contains our solar system. Approximately 5,000 light years from us, far beyond the stars which outline the constellation Perseus, astronomers have identified an adjacent Milky Way arm located towards the Galaxy's outskirts. In the opposite direction, halfway across the sky in the constellation Sagittarius, another neighboring arm is observed. The Sagittarius spiral arm is positioned in the direction of the Galactic center and is seen as a glowing band, the blended light of its distant stars and nebulae. Although it is at about the same distance from us, the Milky Way in Sagittarius appears much broader and brighter than its counterpart in Perseus since there are far more stars and nebulae in the middle of the Galaxy than at its fringes. On evenings in early October, you can trace the tapered shape of the Milky Way's Band as you gaze from Sagittarius low in the southeastern sky towards Perseus rising in the northeast.

During 1953 astronomer William W. Morgan and his colleagues found concentrations of blue supergiant stars in Perseus, at distances of 4,000 to 6,000 light years, indicating the presence of a Galactic spiral arm at that location. Blue supergiant stars provide a means of defining the position of spiral arms in galaxies because their relatively brief lifespans preclude migration far from their birthplaces in the concentrated hydrogen and dust clouds of the arms. Blue supergiants are therefore seen only in or very near spiral arms in galaxies.

Analogous distributions of spectral type B and A stars within about 2,000 light years of the sun have provided a rough outline for the Galactic arm in which we are located. An example of one of these marker stars is Zeta Persei, a third magnitude type B supergiant, located about 1,300 light years from us in the constellation Perseus. Zeta is the most prominent member of a star group called the Zeta Persei Association containing several dozen O and B stars, whose proper motions indicate mutual expansion from a common formation site. This group covers a region of space about 100 light years in diameter containing patches of nebulosity that may be remnants of the cloud out of which these stars were formed.

A similar set of massive, bright, young stars comprises the Alpha Persei Association, seen concentrated in a lovely array around Mirfak, the brightest star in Perseus.

Algol

Northeast **East-Northeast** **East**

Algol AL-gall Beta, β Persei, page 292	2.2 magnitude blue-white B8 main sequence V 105 light-years +0.003 +0.002 +4	−0.2 absolute magnitude 03h 08m +40° 57′ December 24 transit STR 048

Location•This is a well-known variable star, normally of the second magnitude, which is located about one-third of the distance between Mirfak and Hamal.

Meaning•The name Algol is adapted from an Arabian description of this star as the Head of the Demon.

Lore•Although no ancient or medieval references have been found that specifically refer to Algol's light variation, the nature of its early characterizations suggest that the brightness changes had been noted. Throughout the ancient Mediterranean world, Algol was usually associated with some form of dreadful monster. The ancient Israelites are said to have referred to Algol as Lilith, Adam's legendary first wife, who was said to have had frightful characteristics.

In the classical sky legends of Greece and Rome, the star Algol was seen as a representation of the Head of Medusa, one of three loathsome sisters known as Gorgons, who had been punished by the gods for their excessive vanity.

Some of the various descriptions for Algol include the Demon Star, the Head of Medusa, the Eye of Medusa, and the Ghoul Star.

Description•Spectroscopic analysis reveals that Algol consists of three stars that have an angular separation of less than 0.10 arcsecond. The apparent magnitude of the Algol system varies between 2.1 and 3.4 over a period of 2 days, 20 hours, and 49 minutes. This variability is the result of the primary star's periodic eclipse by the larger, cooler secondary.

The primary, designated Algol A, is a blue-white type-B8 mainsequence star, with a luminosity about 150 times greater than that of the sun. The secondary star, which does the eclipsing, is a spectraltype-G8 yellow-white subgiant star, which is approximately six times more luminous than the sun. This star is called Algol B.

Algol A has a surface temperature of about 10,100°K, and its diameter is approximately 3 times that of the sun. The secondary, Algol B, is slightly larger, with a temperature of about 4,900°K. Although these stars are virtually the same size, the greater surface temperature of the primary accounts for its much higher luminosity. The masses of the A and B components of the Algol system have been estimated to be about 3.7 and 0.8 times that of the sun.

The C component is a blue-white main-sequence star with a mass of 1.8 and a luminosity of about 10 times that of the sun.

Components A and B are separated by a distance only 40 percent greater than their individual diameters, and as a result, their mutual gravitation has caused a bridge of gas to extend outwards from the hotter star, filling a gravitational pocket known as a Roche lobe. This lobe of gas has contributed to the characterization of stars similar to Algol as semidetached eclipsing binaries. Studies have indicated that the massive primary is pulling gas away from its subgiant companion's atmosphere.

Epsilon Persei   October 17

Northeast   **East-Northeast**   **East**

Epsilon Persei EP-sigh-lon PURR-see-eye Epsilon, ε Persei, page 292	2.9 magnitude blue-white B0.5 main sequence V 700 light-years +0.017 −0.024 +1	−3.7 absolute magnitude 03h 58m +40° 01' January 5 transit STR 101

Location•Third-magnitude Epsilon Persei is located 9 degrees to the east of Algol.

Lore•Epsilon Persei was shown at the right knee of Perseus in the representation of the constellation contained in Bayer's *Uranometria* of the early seventeenth century.

History•Epsilon Persei has an eighth-magnitude companion star, whose relative position was measured by Wilhelm von Struve in 1832. The pair has the same proper motion, but no evidence of orbital motion has been detected since Struve's time.

Description•The primary star of Epsilon Persei is a blue-white member of the main sequence and is of spectral type B0.5. This star has a surface temperature of about 26,500°K and a luminosity about 2,500 times that of the sun.

Main-sequence stars of this intrinsic brightness are estimated to have masses approximately 6.5 times that of the sun and main-sequence life expectancies of about 30 million years.

The secondary star to Epsilon Persei is also a blue-white, main-sequence object, in this case of spectral type A2. The apparent separation from the primary is slightly less than 9 arcseconds. This companion is rather difficult to see due to the considerable difference in brightness between it and the primary.

Nearby Features

H-H 12: AN HERBIG-HARO OBJECT (4h 29m + 35° 13′)

H-H 12 is a faint patch of nebulosity located about 7 degrees to the southwest of Epsilon Persei. It is an example of an Herbig-Haro object, a molecular cloud feature believed to be caused by the birth of new stars. These objects are named after the American astronomer George H. Herbig of the University of California and Mexican astronomer Guillermo Haro, who first called attention to them.

Recent investigations using infrared and radio telescopes suggest that many, if not all, protostars, during the period before they join the main sequence, emit powerful streams of gas from their surfaces that encounter a disk of dust and gas surrounding the infant star. Some astronomers have suggested that the blast of material from the young star's surface most easily escapes through the polar regions at the top and bottom of this disk, and as a result, two jets of high-velocity gas are formed, flowing in opposite directions. These jets then pick up additional quantities of gas from the surrounding molecular cloud and push them away from the new star. The shock wave produced by this interaction is generally invisible because its light is absorbed by dust in the molecular cloud. However, if the jet's shock wave reaches the surface of a molecular cloud facing earth, the light that it generates becomes visible to us as an Herbig-Haro object.

See the constellation map for Perseus, page 292.

Fomalhaut

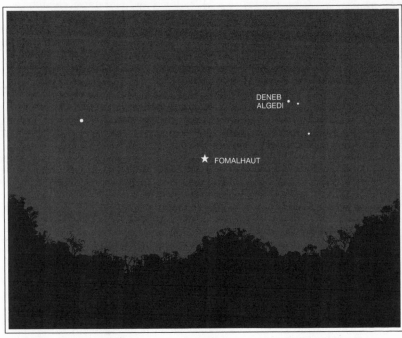

South-Southeast	South	South-Southwest

Fomalhaut
FO-mal-ought
Alpha, α
Piscis Austrini,
   page 302
	1.1 magnitude
	blue-white A3
	main sequence V
	23 light-years
	+0.336 −0.161 +7
		+1.9 absolute
		   magnitude
		22h 58m − 29° 37′
		October 21 transit
		STR 118

Location•Fomalhaut is a prominent first-magnitude star located 21 degrees southeast of Deneb Algedi, in a region removed from other bright stars.

Meaning•In Ptolemy's *Syntaxis* star catalogue, this star is described as "the star in the mouth [of the southern fish]." The name Fomalhaut is adapted from the Arabic translation of Ptolemy's description.

Lore•Fomalhaut—along with Aldebaran, Regulus, and Antares—was considered one of the four Royal Stars, which served as Guardians of Heaven in ancient Persian sky lore. This star has sometimes been portrayed as a part of the water around the feet of Aquarius, and Ptolemy also called it "that at the beginning of the Water." Early

Arabian sky-watchers are said to have called this star the First Frog. Due to the lack of bright stars in its neighborhood, Fomalhaut is sometimes known as the Solitary One.

As one faces south and looks up at the autumn evening sky, it is easy to come face to face with Fomalhaut. Its brightness and isolated position give this star a distinguished identity as a symbol of the autumn season.

With each passing year, as you observe Fomalhaut during the weeks when trees are richly adorned in their fall colors, the title Autumn Star will seem increasingly appropriate for this lovely feature of the southern sky.

Description•Blue-white Fomalhaut is a main-sequence star with an apparent magnitude of 1.1. Its distance from earth is 23 light-years, and it is one of the closer of the bright stars to the earth. Fomalhaut's surface temperature is estimated to be 8,900°K, and this star is about 15 times more luminous than the sun. The spectrum of Fomalhaut is characterized by strong lines of hydrogen and various absorption lines of ionized metals such as iron, titanium, and magnesium.

Main-sequence stars of Fomalhaut's luminosity are estimated to have masses about 1.8 times that of the sun. The main-sequence life expectancy of such stars is in the order of magnitude of 1.5 billion years.

Piscis Austrinus, *the Southern Fish*

DENEB ALGEDI

(CAPRICORNUS)

α FOMALHAUT

ε

δ γ   β   μ   ι

BRIGHT STAR OF PISCIS AUSTRINUS

October 21	Fomalhaut	1.1	blue-white	main sequence

PIE-siss oss-TRY-nus
PsA
Piscis Austrini

Evening Season•October – November

Sky Track•The stars of Piscis Austrinus rise in the southeast, transit the meridian low in the southern sky, and set towards the southwest.

Lore•Piscis Austrinus is one of the many constellations in this part of the sky that is associated with water. Traditional pictures of the constellation figure often show the Southern Fish with its mouth open, drinking a flow of water pouring out of the Water Jar of Aquarius.

Description•Piscis Austrinus is located to the southeast of Capricornus and directly to the south of Aquarius.

This faint constellation is dominated by its one bright star, Fomalhaut, which has an apparent magnitude of 1.1. This constellation's next brightest star is 16 times fainter than this and has a magnitude of only 4.2. For this reason, the elongated figure of stars that marks the Southern Fish is best seen with binoculars.

Diphda

Southeast	South-Southeast	South

Diphda DIF-dah Beta, β Ceti, page 306	2.0 magnitude yellow K1 giant III 60 light-years +0.232 +0.036 +13	+0.7 absolute magnitude 00h 44m −17° 59′ November 17 transit STR 122

Location·Twenty-eight degrees to the northeast of Fomalhaut is located second-magnitude Diphda, the brightest star in the constellation Cetus.

Meaning·The name Diphda was adapted from an early Arabian title for this star, the Second Frog. Fomalhaut in Piscis Austrinus represented the First Frog.

Lore·In the sky lore of Greece and Rome, Diphda marked the southern branch of the Whale's tail; an alternate name, Deneb Kaitos, is from an Arabian translation of this classical description. Astronomers in medieval China knew this star as the Superintendent of the Earthworks.

Description•Diphda is a yellow giant star that has an apparent magnitude of 2.0. At a distance of 60 light-years from earth, this star has a luminosity about 45 times greater than the sun's. With a spectral classification of K1, Diphda has a surface temperature estimated to be approximately 4,350°K.

Absorption lines of neutral metals dominate the spectrum of Diphda. The right ascension and declination coordinates of this star's proper motion are +0.232 and +0.036 arcsecond, respectively. The positive direction of right ascension, indicated by the plus sign, shows an eastward proper motion. The positive direction of declination shows a northward proper motion. Diphda's distance from the sun increases by 13 kilometers every second.

Cetus, *the Whale*

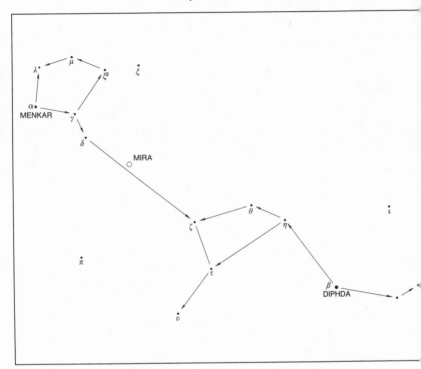

BRIGHT STARS OF CETUS

October 22	Diphda	2.0	yellow	giant
October 27	Menkar	2.5	yellow-orange	giant

SEE-tus
Cet
Ceti

Evening Season•November – January

Sky Track•This is an equatorial constellation. It rises in the east and crosses the meridian halfway between the horizon and the zenith before setting in the west.

Lore•Since at least the time of the Greek writer Aratus in the third century B.C., Cetus has been associated with the sea serpent that was sent to the coast of Ethiopia as punishment to Cassiopeia and that was eventually slain by Perseus as he rescued Andromeda.

Although the title Whale has been applied to this constellation since Roman times, the usual representation of the creature's figure looks more like portrayals of the Loch Ness Monster than of any known cetacean.

A circlet of six stars extending to the northwest from Menkar defines the creature's head, which is separated from its body by a long neck. The smallish body of Cetus is figured from stars including Diphda and some of its eastern neighbors.

Description·Cetus covers a wide portion of the sky and is rather difficult to identify. It straddles the celestial equator to the south of Andromeda and Aries. Cetus is a constellation of mostly faint stars because it lies close to the South Galactic Pole. When we look in the direction of the constellation Cetus, we are viewing a thin section of our galaxy, nearly at 90 degrees to the Band of the Milky Way. This is away from the direction of galactic nebulae such as those in Orion and Scorpius, where groups of massive and highly luminous young stars are seen, and where new generations of brilliant stars are presently being formed.

Alcyone

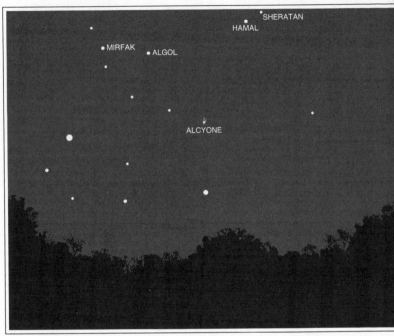

East-Northeast	East	East-Southeast
Alcyone al-SIGH-oh-nee Eta, η Tauri, page 310	2.9 magitude blue-white B7 giant III 400 light-years +0.019 −0.044 +10	−2.6 absolute magnitude 03h 47m +24° 06′ January 2 transit STR 129

Location•Third-magnitude Alcyone is the brightest star in the Pleiades, a lovely open star cluster located in the constellation Taurus. Alcyone is found 16 degrees to the south of the star Epsilon Persei.

Meaning•Alcyone is named after one of the Seven Sisters of the Pleiades, daughters of Atlas and half-sisters of the Hyades, in Greek mythology.

Lore•The stars of the Pleiades form one of the best-known groups in the sky. Richard H. Allen, in his *Star Names: Their Lore and Meaning,* writes that one of the first star descriptions on record is a Chinese reference to the Pleiades, from the year 2357 B.C. During the age of classical Greece, the rise of the Pleiades into the predawn eastern sky coincided with the arrival of fair weather and the start of the navigational season on the Mediterranean. As a result, the Pleiades came to be known as the Sailors' Stars, a title that remained popular in England and Germany up to recent times.

Many lands around the world had special names for the stars of the Pleiades. Anglo-Saxons and medieval Germans called this group the Seven Stars. In Finland and Lithuania the Pleiades were known as a Net, and in Russia, they were called a Sitting Hen. Swedish skywatchers used the description Fur in Frost, a poetic comment on the sparkling delicacy of these stars. In Denmark, the name Evening Hen was used, and in Greenland Eskimos called the Pleiades a Pack of Dogs. Almost the same description, a Close Pack, was used in Wales.

Although the Pleiades are often called the Seven Sisters, only six stars are usually seen distinctly by the average observer. On the other hand, sharp-eyed sky-watchers with excellent observing conditions often report more than seven stars to be visible. Proper-motion studies indicate that the Pleiades cluster contains several hundred member stars.

The Pleiades star cluster was assigned number 45 in Messier's list of celestial objects and is therefore sometimes referred to as M45. It is one of the closest clusters to the earth, with the distance to its center estimated to be about 400 light-years.

Calculations of absolute magnitudes of stars in the Pleiades cluster enable a correlation of these values to be made with star colors on an H-R diagram. The position of the upper left-hand end of a cluster's main sequence suggests to astronomers the approximate time since stars in the cluster formed out of Milky Way dust and gas. The terminus of the most luminous end of the Pleiades' main sequence suggests an age of about 20 million years.

A nebulous haze envelopes stars of the Pleiades and is composed of dust particles that reflect light from the stars. This nebulosity is most evident surrounding the star Merope. Dark absorption lines in the spectrum of this nebula provided the first evidence that some nebulae shine by reflected starlight and not by their own glow.

The Pleiades are a spectacular sight in binoculars, and the brightest stars are seen to resemble a tiny dipper-shaped asterism.

A lovely reference to the Pleiades was made by Alfred, Lord Tennyson, in his poem "Locksley Hall":

Many a night I saw the Pleiads, rising thro' the mellow shade,
Glitter like a swarm of fireflies tangled in a silver braid.

Description•Alcyone, the brightest of the Pleiades stars, has an apparent magnitude of 2.9. It is a blue-white giant that lies at an estimated distance of 400 light-years. This star has a surface temperature of 12,500°K and a luminosity about 900 times greater than that of our sun.

Taurus, *the Bull*

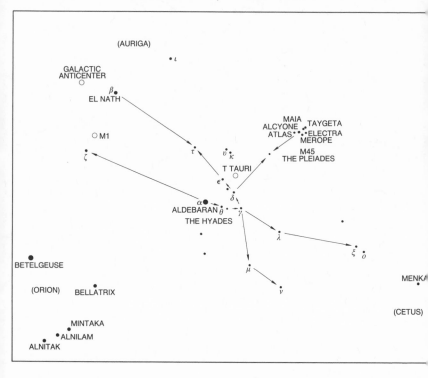

BRIGHT STARS OF TAURUS

October 24	Alcyone	2.9	blue-white	giant
November 10	Aldebaran	0.9	yellow	giant
November 15	El Nath	1.7	blue-white	giant

TAW-rus
Tau
Tauri

Evening Season•November – March

Sky Track•This constellation rises in the northeast, crosses the meridian high in the southern sky, and sets in the northwest.

Lore•Taurus was one of the earliest groups of stars to be considered a constellation. The Bull, a symbol of the vitality of spring, served to mark the sun's position in the sky at the start of spring during the period from about 4000 to 1700 B.C. Since that time, the precessional motion of earth has carried the vernal equinox towards the west, through the constellation Aries and to its present location in the constellation Pisces.

In Greek mythology, Taurus was identified with the Bull whose form Zeus assumed as he tried to woo the maiden Europa. Traditional pictures of Taurus often portray him swimming through the sea with Europa on his back.

Description· Taurus' face is represented by a V-shaped pattern of stars in the Hyades cluster. First-magnitude Aldebaran marks the Bull's fiery eye.

The Hyades open star cluster is centered about 130 light-years from earth, and it is of great importance in the hierarchy of techniques used to measure distances in space. Along with the Ursa Major cluster and the Scorpius-Centaurus cluster, the Hyades is near enough to earth so that direct, trigonometric methods can be used to measure its distance. Distances of more remote open star clusters may be inferred through a comparison of their H-R diagrams with those of the Hyades, Ursa Major, or Scorpius-Centaurus clusters.

The Seven Sisters, or Pleiades, are members of an open star cluster centered about 10 degrees northwest of the Hyades. Stars of the Pleiades resemble a tiny cloud when seen with the naked eye and are a wonderful sight in binoculars or a low-power telescope.

T TAURI: PROTOTYPE FOR A CLASS OF VERY YOUNG STARS
(4h 21m + 19° 29′)

During the past few years, advanced radio, infrared, and ultraviolet observing techniques have enabled astronomers to begin to understand a class of objects known as T Tauri stars.

The irregular variable star designated T Tauri is a tenth-magnitude object located about 5 degrees to the northwest of Aldebaran. This star varies unpredictably in brightness between the ninth and thirteenth magnitude. The associated nebula has brightened considerably since the 1920s, and it appears to be influenced by T Tauri.

During the 1940s, other stars similar to T Tauri were found by Alfred H. Joy of the Mount Wilson Observatory in neighboring regions of Taurus and Auriga, as well as in the nebulae M16 in Serpens, M42 in Orion, and NGC 2264 in the constellation Monoceros. These stars all display erratic light variations of up to three magnitudes, intense spectral-emission lines, and associations with dust-laden molecular clouds. T Tauri stars also show evidence that they are ejecting quantities of their mass into space at high speeds.

When the luminosities and temperatures observed for these stars are plotted on a Hertzsprung-Russell diagram, it appears that these are very young stars, still in the process of gravitational contraction, which have not yet reached the main sequence. These objects are estimated to range in age from 100,000 to 1 million years and in mass from 20 percent to 3 times that of the sun.

Menkar   October 27

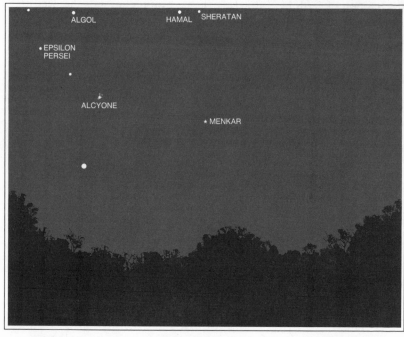

East	East-Southeast	Southeast

Menkar	2.5 magnitude	−0.7 absolute
MEN-kar	yellow-orange M2	magnitude
Alpha, α	giant III	03h 02m +04° 05′
Ceti, page 306	150 light-years	November 6 transit
	−0.012 −0.074 −26	STR 142

Location•Third-magnitude Menkar, in the constellation Cetus, lies at the southern corner of an equilateral triangle whose northern corners are marked by the Pleiades and the star Hamal in Cetus. Twenty-three degrees of arc separates each of these objects.

Meaning•This star represented the Tip of the Nose of Cetus in Ptolemy's description of the constellation. The present name is a variation of an Arabic translation of the Greek designation.

Lore•Some traditional star maps showed Menkar at the open mouth of the Whale–Sea Monster figure of Cetus. The closed figure of six stars of which Menkar is a part marks the head of Cetus in Greek and Roman sky lore. This same rough circle of stars represented a hand of the Pleiades in early Arabian sky legends and also Heaven's Round Granary in the astronomical lore of ancient China.

Description•Menkar is a yellow-orange giant star whose luminosity is approximately 150 times that of the sun. The surface temperature of a star with this type of spectrum is estimated to be 3,250°K. The comparison of Menkar's apparent magnitude of 2.5 with its estimated luminosity indicates a distance of about 150 light-years from earth.

Nearby Feature

MIRA, "THE WONDERFUL": LONG-PERIOD VARIABLE STAR
(2h 19m −2° 59′)

Mira is the prototype for a class of long-period pulsating variable stars and was first identified to be variable by the German astronomer David Fabricius in 1594. This orange star is located 15 degrees southwest of Menkar and has one of the coolest surfaces known for a star—between 1,900°K and 2,400°K. It is a supergiant whose diameter was measured by interferometer to be approximately 420 times that of the sun. In the size scale of the solar system, at its average diameter Mira would extend outwards to halfway between the asteroid belt and Jupiter's orbit, and when it swells to maximum size it would nearly reach Jupiter's orbital distance.

This great star, which is near the end of its life cycle, varies in brightness between magnitudes of about 3 to 4, at its maximum, to an invisible-to-the-naked-eye magnitude 9 at minimum light over a period of about 331 days.

With surface temperatures that vary between about 1,900°K and 2,400°K, molecules such as titanium oxide and even water vapor have been detected in the outer atmosphere of Mira.

The Bright Stars of Autumn

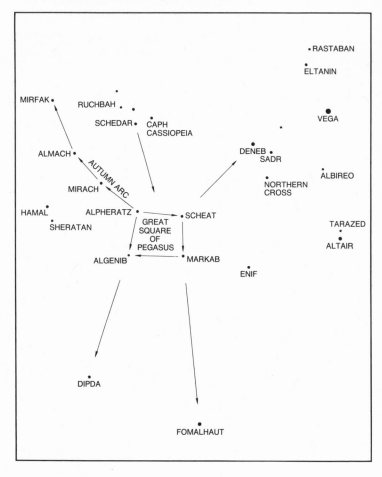

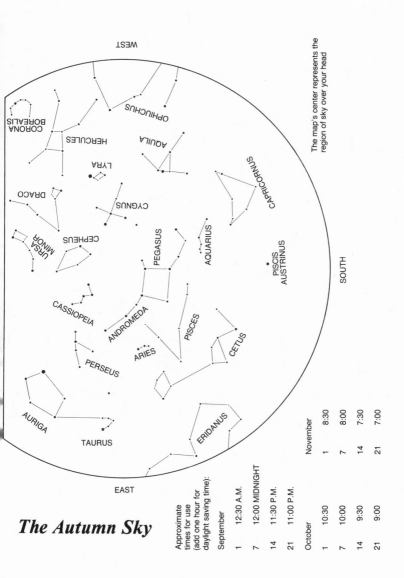

WEST

The map's center represents the
region of sky over your head

CORONA
BOREALIS

OPHIUCHUS

HERCULES

AQUILA

LYRA

DRACO

CYGNUS

CAPRICORNUS

URSA
MINOR

CEPHEUS

PEGASUS

AQUARIUS

SOUTH

PISCIS
AUSTRINUS

CASSIOPEIA

ANDROMEDA

PISCES

CETUS

PERSEUS

ARIES

AURIGA

ERIDANUS

TAURUS

EAST

SOUTH

The Autumn Sky

Approximate
times for use
(add one hour for
daylight saving time):

September

1	12:30 A.M.
7	12:00 MIDNIGHT
14	11:30 P.M.
21	11:00 P.M.

October

1	10:30
7	10:00
14	9:30
21	9:00

November

1	8:30
7	8:00
14	7:30
21	7:00

November Stars

Capella   November 2

North-Northeast	Northeast	East-Northeast
Capella kah-PELL-ah Alpha, α Aurigae, page 318	0.1/0.6 magnitude white G0 giant III 46 light-years +0.080 −0.423 +30	−0.1 absolute magnitude 05h 17m +46° 00′ January 25 transit STR 165

Location•Capella is a brilliant star located in the constellation Auriga, the Charioteer. It is found about 20 degrees to the east of Mirfak.

Meaning•The name Capella comes from Rome, where it characterized this as the Goat Star.

Lore•In Mesopotamia, Capella was known as the Leader since it was the brightest star in the vicinity of the sun at the vernal equinox at the start of the new year.

In the sky lore of Greece and Rome, Capella was often associated with rain and storms since it was visible in the predawn sky during springtime. During classical times, Capella was also sometimes associated with a cornucopia or a broken goat's horn.

Ptolemy described Capella as marking Auriga's left shoulder.

History•In 1899, spectroscopic analysis of Capella performed by William Wallace Campbell at Lick Observatory revealed this star to be a spectroscopic binary. During 1920, John A. Anderson used the interferometer on the 100-inch telescope at Mount Wilson to help determine the orbital characteristics of the Capella binary system; it indicated a separation of about 0.05 arcsecond. This was the first use of the interferometer for an orbit determination.

Description•Capella is the fourth-brightest star in our night sky. With an apparent combined magnitude of 0.1, this binary star is more than twice as bright as first-magnitude stars such as Spica and Antares, and its glory dominates the northern part of the evening sky from autumn through winter and well into springtime. Due to its high northern declination, Capella is visible at some time during the night throughout the entire year.

The diamondlike brilliance of Capella is blended light of a pair of nearly identical giant stars, which orbit their common center of gravity with a separation approximately equal to the distance between the sun and Venus in our solar system. The components of Capella have an orbital period of just over 104 days.

The larger component of Capella has a mass, determined from orbital calculations, 2.67 times that of the sun. This star has a white color and is 100 times more luminous than the sun. Its surface temperature is estimated to be about 5,000°K, and its diameter is approximately 10 times larger than the diameter of the sun.

The second of the twin stars in the Capella system has a mass estimated to be 2.55 times that of the sun. Its spectrum indicates a yellow-white color, and its luminosity is about 60 times that of the sun. Stars of this type are believed to have surface temperatures of about 5,600°K. This less massive star has only about 60 percent of its neighbor's luminosity because its diameter is but 6 times that of the sun and it therefore has less surface area from which to shine light into space than does its larger companion.

A third component, known as Capella H, is a yellow-orange spectroscopic binary that has a common proper motion with the stars described above. The components of this double are dwarf stars with an apparent magnitude of about 10 and a separation from the bright stars of Capella of 723 arcseconds. This angular distance is equal to about one-third the diameter of the full moon.

Auriga, *the Charioteer*

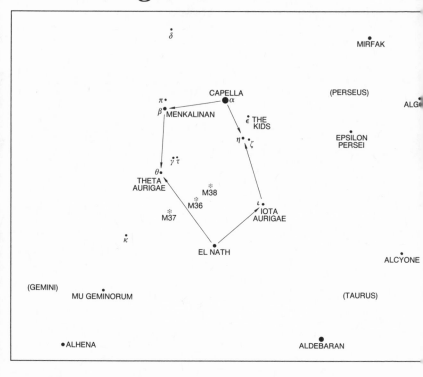

BRIGHT STARS OF AURIGA

November 2	Capella	0.1	white-yellow	giant
November 5	Iota Aurigae	2.7	yellow	giant
November 14	Menkalinan	1.9	blue-white	subgiant
November 19	Theta Aurigae	2.6	blue-white	peculiar

or-EYE-gah
Aur
Aurigae

Evening Season•November – April

Sky Track•This constellation rises from the northeast, passes overhead, and sets in the northwest.

Lore•The constellation figure of Auriga probably originated in the ancient Middle East.

In Greek legends, Auriga has been associated with Erichthonius, son of Hephaestus, the god of fire, and Mother Earth. Erichthonius was a mythical early king of Athens who was lame and was said to have invented the chariot as a means of transportation.

Description • Auriga lies across the centerline of the Milky Way. It contains several open star clusters visible with a small telescope.

The brightest star in the constellation, Capella, is also known as the Goat Star. Three nearby stars, Eta, Epsilon, and Zeta, form a small triangle and are known as the Kids.

El Nath, the star at the southeastern corner of the Pentagon of Auriga, is formally assigned to the constellation Taurus, and it marks the tip of one of the Bull's horns.

Nearby Feature

EPSILON AURIGAE: AN UNUSUAL ECLIPSING SYSTEM
(5h 02m +43° 49′)

Epsilon Aurigae is a third-magnitude star that may be found just 3 degrees to the southwest of Capella. It is at the northern tip of a small triangle of stars that marks the figure known as the Kids of Auriga.

Epsilon is a blue-white supergiant estimated to be 3,500 light-years from earth. The star's brightness, diameter, and spectrum vary slightly over a period of about 110 days, and this star seems to be surrounded by a shell of gas, which is revealed by spectral emission lines.

The remarkable characteristic of Epsilon Aurigae, which causes it to be extremely interesting to astronomers, is an eclipse that reduces the star's apparent magnitude to 3.8 once every 27 years. It is speculated that an unseen but enormous object slowly moves along an orbit to a position in front of Epsilon, reducing the light that reaches earth to about 50 percent of the star's normal brightness. It requires about half a year for the light to fade, and minimum brightness is maintained for one year. Restoration of the star's brightness to magnitude 3.0 takes an additional half-year. The extraordinary duration of the eclipse indicates that the eclipsing object is of exceptionally large size.

Astronomers have proposed several models to explain the long eclipse of Epsilon Aurigae, which last occurred in 1982. For many decades the standard explanation has been that the eclipsing object is a gigantic star, perhaps 3,000 times the size of the sun.

Recent searches using modern infrared telescopes have, however, failed to detect infrared radiation from Epsilon Aurigae's companion, and the super-supergiant theory may therefore have to be put aside and alternative theories considered. Some of these suggest that the eclipsing body may be a huge shell or disk of gas and dust that might surround a comparatively faint star whose light is lost in the glare from Epsilon Aurigae.

Iota Aurigae   November 5

Northeast	East-Northeast	East

| Iota Aurigae
EYE-oh-tah or-
EYE-jay
Iota, ι
Aurigae, page 318 | 2.7 magnitude
yellow K3
giant II
250 light-years
+0.004 −0.018 +18 | −2.3 absolute
magnitude
04h 57m +33° 10m
January 20 transit
STR 177 |

Location•Iota Aurigae is located at the southwestern corner of the pentagon-shaped asterism formed by bright stars of the constellation Auriga. Iota is about 12 degrees to the south of Capella, and it is a third-magnitude star.

Description•This star has an apparent magnitude of 2.7 and is a yellow giant at a distance estimated to be about 250 light-years from earth. Stars of this spectral and luminosity type are about 700 times more luminous than our sun. Iota Aurigae has a temperature of about 4,000°K.

Absorption lines of neutral metals are the strongest lines seen in the spectrum of Iota Aurigae. The annual proper motion of this star is

0.004 arcsecond to the east and 0.018 arcsecond to the south. The combined proper motion is 0.018, almost directly south on the celestial sphere. At this rate, Iota Aurigae would require about 100,000 years to traverse a distance on the celestial sphere equal to the apparent diameter of the full moon.

The radial velocity of Iota, obtained through measurements of the Doppler shifts of its spectral lines, is equal to + 18 kilometers per second. The plus sign indicates that the distance between Iota Aurigae and the sun is increasing.

Aldebaran   November 10

East-Northeast   **East**   **East-Southeast**

Aldebaran al-DEB-ah-ran Alpha, α Tauri, page 310	0.9 magnitude yellow K5 giant III 70 light-years +0.065 −0.189 +54	−0.7 absolute magnitude 04h 36m +16° 31′ January 15 transit STR 196

Location•Aldebaran continues the rich progression of first-magnitude stars that enters the evening sky at this time of the year. It is located 16 degrees to the south of Iota Aurigae and 14 degrees to the south of Alcyone.

Meaning•The origin of Aldebaran's name is uncertain, although it has been suggested that it was taken from a description meaning the Driver or Follower of the Pleiades.

Lore•Aldebaran is the ninth-brightest star in our sky, and it was celebrated in ancient times as one of the four Royal Stars, which marked the seasons of the year. (The other Royal Stars are Regulus, Antares, and Fomalhaut.)

Taurus marked the vernal equinox and the start of the year around 3000 B.C., and as a result, Aldebaran was known as the Leading Star of Stars. It was also known in Mesopotamia as the Messenger of Light, a title also applied at various times to Capella and Hamal, as the process of precession relocated the site of the vernal equinox.

Ptolemy described Aldebaran as the Bright Red Star in the Hyades at the southern eye of the Bull. His description as red was probably a poetic exaggeration prompted by Aldebaran's obviously different color compared to its neighbors.

Aldebaran's most popular description has been the Eye of the Bull. This animal's face is outlined by the V-shaped asterism formed by stars in the Hyades cluster.

History• An occultation of Aldebaran by the moon was observed in Athens in the year A.D. 509 and its characteristics recorded. According to Robert Burnham, this observation provided a clue that suggested to Edmund Halley Aldebaran had changed its position slightly by the seventeenth century. Halley then checked the locations of several other bright stars, including Procyon and Sirius, and concluded that they also had moved by several minutes of arc. Such changes of star positions are now known as proper motions.

Description• Aldebaran is a giant star with a diameter measured with the interferometer to be 45 times that of the sun, a size equivalent to half the diameter of the planet Mercury's orbit. This star is 70 light-years from earth, and it shines in our sky with an apparent magnitude of 0.9. Knowledge of Aldebaran's distance and apparent magnitude enables us to calculate its luminosity to be 150 times greater than that of the sun.

In keeping with romantic descriptions of this star as the fiery eye of Taurus the Bull, Aldebaran's surface temperature of 3,800°K indicates a yellow color.

Infrared observations suggest that Aldebaran is surrounded by a relatively cool shell of gas, and, as is the case with stars of this type, it is slightly variable in brightness, ranging between magnitudes 0.75 and 0.95. A yellow-orange dwarf star with a magnitude of 13.5 is a binary companion, separated from Aldebaran by 31 arcseconds.

Nearby Feature

THETA[1] AND THETA[2] TAURI (4h 29m + 15° 58′ AND 4h 29m + 15° 52′)

The star Theta[2] Tauri is the brightest member of the Hyades cluster and it forms a lovely binocular double with its companion, Theta[1] Tauri. The separation of this pair is 337 arcseconds and it is located about 2 degrees to the southwest of Aldebaran. Theta[2] is blue-white with an apparent magnitude of 3.4, while Theta[1] has a magnitude of 3.8 and a yellow color.

Menkalinan   November 14

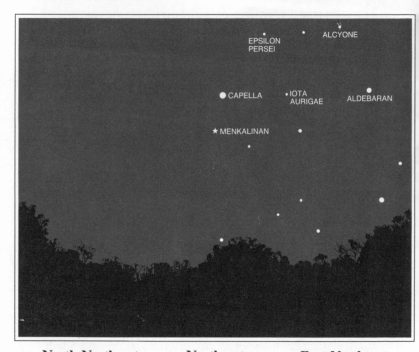

North-Northeast	Northeast	East-Northeast
Menkalinan men-KAL-ih-nan Beta, β Aurigae, page 318	1.9/2.7 magnitude blue-white A2 subgiant IV 90 light-years −0.055 −0.001 −18	−0.5 absolute magnitude 06h 00m +44° 57′ February 5 transit STR 210

Location·Menkalinan is a second-magnitude star situated at the northeastern corner of the pentagon-shaped asterism of Auriga. It is the second-brightest star in this constellation and lies 7 degrees to the east of brilliant Capella.

Meaning·The name Menkalinan was adapted from the Arabian translation of this star's Greek description as the Star in the Right Shoulder of the Charioteer.

History·Antonia C. Maury of the Harvard College Observatory found as a result of observations that she made in 1889 that Menkalinan is a spectroscopic binary.

In 1910 Joel Stebbins, a pioneer in the development of photoelectric photometry, used one of his electric light-measuring devices at the Washburn Observatory of the University of Wisconsin to study variations in Menkalinan's brightness. His analysis of the star's magnitude variations showed that Menkalinan is part of a short-period eclipsing-binary system. Stebbins was able to determine this star system's orbital characteristics as a result of his photometric research.

Description•The Menkalinan system consists of two virtually identical blue-white subgiant stars in a nearly circular orbit inclined about 12 degrees from our line of sight. As a result of this small inclination, the stars partially eclipse each other as they orbit their common center of gravity. This partial eclipse results in a drop in brightness of about 8 percent once every 3 days, 23 hours, and 3 minutes.

Both components have spectral classifications of A2, and the analysis of their orbit indicates that their masses are 2.33 and 2.25 times that of the sun. Each component has a diameter about 3 times that of our sun, a luminosity 50 times that of our sun, and a surface temperature of 9,200°K.

Menkalinan's distance from the earth is revealed by its trigonometric parallax to be 90 light-years, and its space motion suggests that it belongs to the Ursa Major star cluster. This is the nearest star cluster to us, and it includes five of the seven stars in the Big Dipper.

Menkalinan is located very near the 6-hour line of right ascension, which is part of the solstitial colure. This great circle intersects the ecliptic at the point of the summer solstice, the sun's highest position in the sky for observers in the Northern Hemisphere.

El Nath   November 15

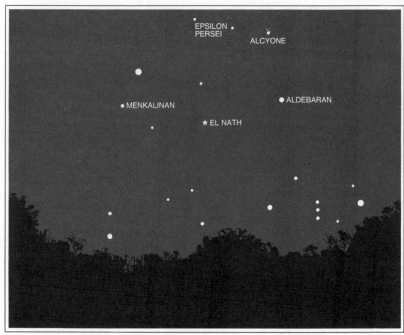

Northeast	East-Northeast	East

El Nath	1.7 magnitude	−2.0 absolute
EL nath	blue-white B2	magnitude
Beta, β	giant III	5h 26m + 28° 36′
Tauri, page 310	200 light-years	January 27 transit
	+0.025 −0.175 +9	STR 215

Location•El Nath is found at the southern tip of the pentagon of Auriga. It is a second-magnitude object 17 degrees to the south of Capella and nearly as distant from Aldebaran, which lies to the southwest of El Nath.

Meaning•The name El Nath was apparently applied to this star around the beginning of the nineteenth century. It originally had been a name for the star Hamal in the constellation Aries and is derived from an Arabic description meaning the Butting One.

Lore•El Nath marks the northern horn tip of Taurus, the Bull. In the past, this star was also considered to be a part of Auriga, where it served to represent the Charioteer's heel.

In ancient India this star was associated with the god of fire.

History•In 1928, the International Astronomical Union formally defined boundaries for all the constellations, with El Nath officially assigned to Taurus.

Description•El Nath is a blue-white giant star estimated to be about 200 light-years from earth. Its surface temperature is approximately 12,500°K, and it is 550 times more luminous than the sun.

The space velocity of El Nath suggests that it may be an outlying member of the Pleiades star cluster.

Nearby Feature

M1, THE CRAB NEBULA: A SUPERNOVA REMNANT
(5h 33m +22° 05′)

On July 4, 1054, Chinese astronomers recorded the sudden appearance of what they called a "guest star" in the constellation Taurus. The object was as bright as the planet Venus, and for three weeks it could be seen even in the daytime. After about twenty-one months the star faded to invisibility. Although no European records of this event have yet surfaced, pictographs drawn on rocks in northern Arizona suggest that the guest star was also seen and recorded by American Indian observers.

In 1731, an English amateur astronomer named John Bevis discovered a wisp of nebulosity about 6.5 degrees to the south of El Nath and just 1 degree northwest of the third-magnitude star Zeta Tauri, which marks the Bull's southern horn tip. Charles Messier, in 1758, also noted this object and assigned it the first number in his list of nebulae and star clusters. By 1844, Lord Rosse in Ireland, using his mammoth 72-inch reflecting telescope, discovered filaments emanating from the nebula's center that reminded him of crab legs and inspired the object's name.

The Crab Nebula, M1, appears as a hazy oval of light in a 6-inch telescope, and photographs taken with large telescopes are needed to reveal the Crab's elaborate structure. The distance to the Crab Nebula is estimated to be 6,500 light-years.

This is the best-known example of a supernova remnant. The nebula represents the outer layers of the massive star whose explosion became visible on earth in 1054. This supergiant's core collapsed to become an object with a diameter of about 20 kilometers and a density equal to that in an atomic nucleus. Such a superdense object had been postulated since the 1930s by various astronomers as a consequence of the explosion of very massive stars. These objects came to be called neutron stars, since their internal pressures would be so great that protons and electrons would be squeezed together to form neutrons.

Theta Aurigae   November 19

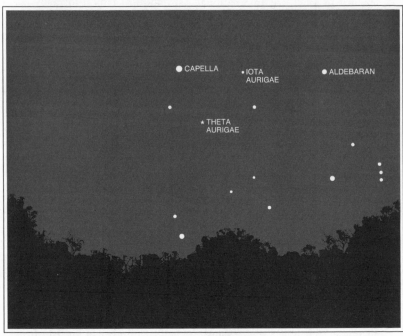

Northeast   **East-Northeast**   **East**

Theta Aurigae	2.6 magnitude	−0.7 absolute
THAY-tah	blue-white A0p	magnitude
or-EYE-jay	peculiar	06h 00m +37° 13′
Theta, θ	150 light-years	February 5 transit
Aurigae, page 318	+0.049 −0.082 +30	STR 230

Location•Theta Aurigae is on the eastern side of the pentagon of Auriga. It is a third-magnitude star located 7 degrees to the south of Menkalinan.

Description•This star has an apparent magnitude of 2.6 and it shines with a blue-white color. The spectrum of Theta Aurigae is classified as peculiar due to unusually prominent absorption lines of the element silicon. Theta is sometimes called a silicon star because of the presence of these spectral lines.

At a distance of about 150 light-years from earth, Theta Aurigae is estimated to have a luminosity about 150 times that of the sun. The surface temperature of this star is approximately 9,900°K.

Theta has a seventh-magnitude companion at a distance of 4.5 arcseconds.

Nearby Feature

M37: OPEN STAR CLUSTER (5h 51m +32° 32′)

There are three attractive open star clusters in the constellation Auriga: M36, M37, and M38, with M37 considered to be the finest of the lot.

M37 resembles a small sixth-magnitude nebula when seen with binoculars. It is located approximately 5 degrees to the south of Theta Aurigae. When observed through telescopes 3 inches or more in diameter, M37 begins to be resolved into a marvelous field of sparkling stars.

This open star cluster is believed to contain at least 400 stars, and the analysis of the upper termination point of its main sequence suggests that M37 was formed about 500 million years ago. Its distance is estimated at 4,500 light-years from the earth.

See the Auriga constellation map, page 318.

Bellatrix

December 1

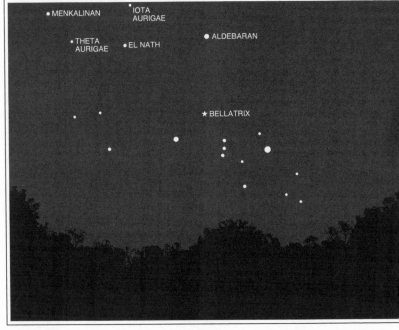

East	East-Southeast	Southeast
Bellatrix beh-LAY-trix Gamma, γ Orionis, page 332	1.6 magnitude blue-white B2 giant III 300 light-years −0.012 −0.014 +18	−3.3 absolute magnitude 05h 25m +06° 21′ January 27 transit STR 279

Location•Second-magnitude Bellatrix is located 16 degrees to the southeast of Aldebaran. It is the first of the bright stars in the magnificent constellation Orion to rise in the eastern sky.

Meaning•The derivation of the name Bellatrix is obscure. It is a Latin word meaning "female warrior."

Lore•This star's Arabian title meant "the leader," because it is the first of Orion's bright stars to rise in the sky.

Description•Bellatrix has an apparent magnitude of 1.6 and shines with a blue-white color, characteristic of its surface temperature of 23,000°K. Its distance from earth is estimated to be about 300 light-years, and its luminosity is 1,800 times that of the sun.

Bellatrix appears to be surrounded by an expanding shell of gas, and there are indications that this star is ejecting quantities of its atmosphere out into space. It is a member of the Orion OB1 Association of massive and comparatively young stars. Recent studies indicate that stars in this association are less than 10 million years old and the youngest of these are in the area of the Great Nebula in Orion.

Other prominent members of the Orion OB1 Association include the three stars of Orion's Belt and the star Saiph, in the southeast corner of the figure. Observations made with Columbia University's 1.2-meter radio telescope led to the discovery of a vast complex of molecular clouds in this region of the sky. This cloud system probably provided the materials out of which many of the stars of Orion were formed. The Orion nebula represents a "hot spot" on these molecular clouds where stars are currently being formed.

Behind the Orion nebula, along our line of sight, lie the cold reaches of the Orion Molecular Cloud Complex. These vast clouds contain hydrogen, helium, and other gases as well as large quantities of interstellar molecular dust. The temperatures of these clouds are estimated to average about 20°K, a level at which complex molecules survive, shielded by the dust particles of the clouds from disruption by ultra violet radiation.

Carbon monoxide molecules found within these clouds produce a characteristic radio wavelength that enables astronomers to trace the carbon monoxide and thereby the extent of the molecular clouds. The great molecular clouds in Orion have been found to stretch across the constellation figure from the star Saiph, northwest through the Sword, and then to the northeast, beyond Alnitak in Orion's Belt.

The Orion molecular clouds appear to have diameters of about 100 to 200 light-years and to contain material equal to the masses of over 100,000 suns. These clouds contain gas and dust that may someday condense to form new stars.

As stars, dust, and gas revolve around the center of the galaxy, they periodically encounter regions defined by density waves. In regions of compression, star formation is stimulated within the dust and gas clouds. The brilliant young stars and luminous gas clouds define the spiral arm system of the galaxy. This "density wave" model of spiral arm formation was devised by C. C. Lin, Frank Shu, and Chi Yuan in 1969. In the neighborhood of the sun, the most active sites of current star formation are found in the direction of Taurus, Auriga, Ophiuchus, and most spectacularly in the constellation Orion.

Orion, *the Hunter*

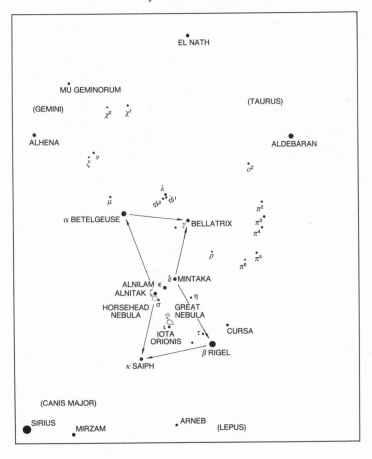

BRIGHT STARS OF ORION

December 1	Bellatrix	1.6	blue-white	giant
December 8	Betelgeuse	0.8	yellow-orange	supergiant
December 9	Mintaka	2.2	blue-white	giant
December 10	Alnilam	1.7	blue-white	supergiant
December 13	Alnitak	1.8	blue-white	supergiant
December 14	Rigel	0.1	blue-white	supergiant
December 16	Iota Orionis	2.8	blue-white	giant
December 24	Saiph	2.1	blue-white	supergiant

oh-RYE-un
Ori
Orionis

Evening Season•December – March

Sky Track•Orion is an equatorial constellation that is bisected by the celestial equator. It rises at the eastern point of the horizon, crosses the meridian halfway between the horizon and the zenith, and sets directly to the west.

Lore•The derivation of the name Orion is veiled in antiquity. It might come in part from a Mesopotamian god called Uru-anna, who was known as the Light of Heaven.

Ancient Israelites saw these stars as the Giant, Nimrod; and until about 500 B.C., Orion was known in Greece simply as the Warrior. By the time of Euripides, a generation later, the name Orion had come into general use.

The predawn rise of Orion coincides with the peak of summer's heat. The constellation's midnight rise takes place when forests in the north temperate zone show their glorious color changes. In December, Orion makes its debut into the evening sky at the start of the frosty holiday season.

Greek mythology describes Orion as a prince of Boeotia, an ancient kingdom northwest of Athens. Orion's birth was a reward to his father, King Hyrieus, for his hospitality to the gods Zeus, Hermes, and Poseidon. Legends tell us that Orion grew to such a great size that he could walk on the floor of the sea without wetting his head. He became very strong and a mighty hunter, and was followed everywhere by his faithful hunting dogs.

After many adventures, Orion traveled to Crete where he befriended Diana, goddess of the moon and of the hunt. They fell in love and she spent much time with him on Crete. As a result, Diana neglected her responsibility for driving the lunar chariot across the night sky. Her brother Apollo, driver of the solar chariot, warned Diana to return to her duties, but she would not leave Orion. In time, Apollo tricked her into killing Orion with an arrow as he swam far from shore. After this tragedy, Diana transported Orion into the sky as the most glorious of the constellations.

Description•The brilliant stars of Orion are features of a section of the local Milky Way spiral arm, sometimes known as the Orion Spur. Orion is the only constellation in our sky that displays two first-magnitude stars, Betelgeuse and Rigel. These offer the best example of color contrast visible to the naked eye: Betelgeuse is called a red giant, but it actually appears as yellow-orange; Rigel, diagonally across the constellation figure from Betelgeuse, is a blue-white star.

Mu Geminorum   December 3

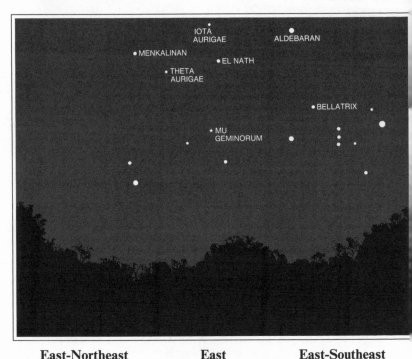

East-Northeast	East	East-Southeast
Mu Geminorum mew gem-in-OAR- um Mu, μ Geminorum, page 336	2.9 magnitude yellow-orange M3 giant III 150 light-years +0.055 −0.112 +55	−0.4 absolute magnitude 06h 23m +22° 31′ February 11 transit STR 287

Location•Mu Geminorum is found 15 degrees southeast of El Nath. It is a third-magnitude star.

Lore•This star was described in Tycho Brahe's star catalogue as the Heel of Castor, the Twin.

Description•Mu Geminorum has an apparent magnitude of 2.9 and is a yellow-orange giant star. Its distance from the earth is about 150 light-years, and it has a luminosity about 100 times greater than our sun's intrinsic brightness. The estimated surface temperature is 3,100°K, which is near the lower end of the stellar range.

As is the case with many giant stars of this spectral class, Mu Geminorum is an irregular variable. Its magnitude range is between 2.76 and 3.02.

Nearby Features·See the following constellation map.

THE SUMMER SOLSTICE (6h 00m +23° 24′)

The sun reaches the most northerly point on the ecliptic when it is about 5 degrees to the west of Mu Geminorum. This point is known as the summer solstice, and it is defined on the celestial sphere by the intersection of the 6-hour coordinate of right ascension with the ecliptic.

M35: OPEN STAR CLUSTER (6h 08m +24° 21′)

The star cluster designated M35 has a position 3.5 degrees to the northwest of Mu Geminorum and 2 degrees northeast of the imaginary point of the summer solstice. This is a fine open cluster for observers using small telescopes and is even visible with binoculars.

M35 contains nearly 500 stars, distributed over a volume of space about 30 light-years in diameter. From its distance of about 2,800 light-years, M35 appears to cover an area of sky equal to the size of the full moon's disk. A low-power eyepiece is needed in order to fit M35 into a telescope's field of view.

HUBBLE'S VARIABLE NEBULA: NGC 2261 (6h 38m +8° 45′)

This strange, triangular nebulosity extends northwards like a fan from the variable star R Monocerotis. It is located about 15 degrees to the south of Mu Geminorum.

The nebula was discovered by William Herschel in 1783. Edwin Hubble found it to vary in size, brightness, and structure as a result of photographic comparisons made in 1916.

Observations made during recent decades suggest that the star associated with Hubble's Variable Nebula is either a T Tauri–type variable or perhaps a protostar at an even earlier stage of its gravitational contraction.

THE ROSETTE NEBULAE (6h 30m +4° 54′)

The Rosette Nebula is a doughnut-shaped cloud of glowing gas that covers an area of sky about three times the size of the full moon's disk, 20 degrees to the south of Mu Geminorum. The light from this nebula is faint and it is barely visible with binoculars even under the best conditions. Long-exposure photographs, however, reveal the Rosette to be a beautiful and complicated structure.

At a distance estimated to be about 2,500 light-years, the Rosette Nebula's diameter is about 60 light-years. It encompasses the extremely young open star cluster NGC 2244, which is located in the open region at the center of the Rosette Nebula. The formation of these stars evidently consumed considerable amounts of dust and gas in this central zone, and then stellar winds from the young stars later blew away most of the remaining nebulosity, leaving the middle of the Rosette relatively clear of interstellar material.

Gemini, *the Twins*

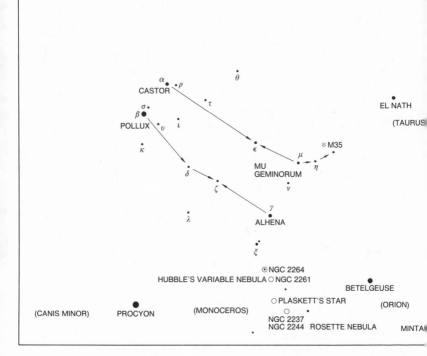

BRIGHT STARS OF GEMINI

December 3	Mu Geminorum	2.9	yellow-orange	giant
December 12	Alhena	1.9	blue-white	subgiant
December 17	Castor	1.6	blue-white	main-sequence
December 21	Pollux	1.2	yellow	giant

GEM-in-ee
Gem
Geminorum

Evening Season•December – May

Sky Track•Gemini rises in the northeast, transits the meridian high in the southern sky, and sets in the northwest during the warm evenings of late spring.

Lore•Castor and Pollux, the prime features of this constellation, form a striking pair and have been known as twin stars in many lands.

Castor and Pollux are Greek names for the sons of Leda. Zeus, under the guise of a swan, came to Leda and was the father of immortal Pollux, the boxer. King Tyndareus was the father of mortal Castor, the horseman. Together the brothers were sometimes called the Dioscuri and both considered to be sons of Zeus. Castor and Pollux were members of the argonautic expedition that sought the Golden Fleece.

The Twins of Gemini were considered patrons of navigators and warriors. They were carved into figureheads for ships and served as symbols of the Roman armies. In the ancient Roman Forum, the temple of Castor and Pollux was a major structure, and today its striking ruins stand as a tribute to these two symbols of vitality and hope.

In Israel, they were said to represent Simeon and Levi, the Brethren.

Description•Gemini lies to the northeast of Orion. The stars of Gemini are on the eastern fringe of the Milky Way and mark the location of the sun during the first weeks of summer.

Two lines of stars extend westward from Castor and Pollux, and these may be imagined to represent the bodies of the Twins.

Nearby Feature

PLASKETT'S STAR: THE MOST MASSIVE BINARY SYSTEM KNOWN (6h 37m +6° 08′)

Plaskett's Star is located 10 degrees to the south of Alhena. It is a spectroscopic binary star at a distance of approximately 2,500 light-years from earth and with an apparent magnitude of 6.1, which is near the threshold of naked-eye visibility.

This system contains two of the most massive stars known. The components of Plaskett's Star each are estimated to contain 55 times the mass of the sun.

The system has an orbital period of 14.4 days, and the stars seem to be surrounded by a region of circumstellar dust. The mammoth components of Plaskett's Star belong to the Monocerous OB Association and are probably related to star cluster NGC 2264.

Cursa

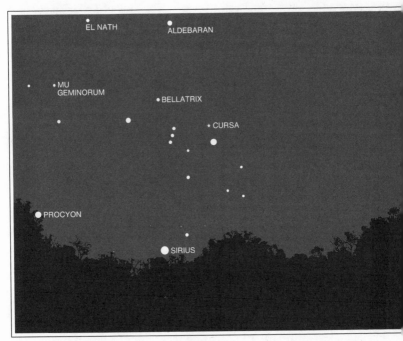

EL NATH

ALDEBARAN

MU
GEMINORUM

BELLATRIX

CURSA

PROCYON

SIRIUS

East-Southeast	Southeast	South-Southeast
Cursa KUR-sah Beta, β Eridani, page 340	2.8 magnitude blue-white A3 giant III 65 light-years −0.100 −0.080 −9	+1.3 absolute magnitude 05h 08m −05° 05′ January 23 transit STR 304

Location•Third-magnitude Cursa is located 3.5 degrees to the north-west of the brilliant star Rigel in Orion.

Meaning•Cursa, along with its neighbors Lambda and Omega Eridani, was known in Arabia as the Footstool of Orion and its name was adapted from this title.

Lore•This star is nearly at the eastern end of the long, winding constellation Eridanus, the River.

Before the influence of Greek astronomy became prevalent, Cursa and its two neighbors were known in the Arabian desert as the Ostrich Nest.

It is said that in ancient China, Cursa was called the Golden Well.

Description•Cursa shines with an apparent magnitude of 2.8 and it is a blue-white giant star. Its spectrum indicates a surface temperature of about 8,900°K. The trigonometric parallax of Cursa is 0.02 arcsecond, which indicates a distance of 20 parsecs, or 65 light-years. At this distance the star's apparent magnitude corresponds to a luminosity 25 times greater than that of the sun.

Eridanus, *the River*

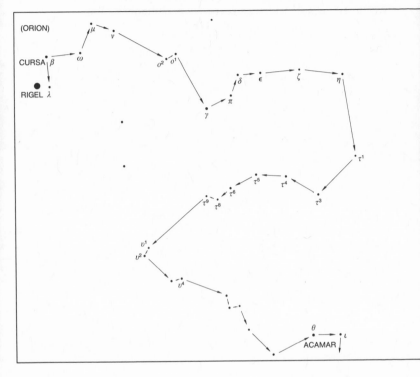

BRIGHT STAR OF ERIDANUS

December 7	Cursa	2.8	blue-white	giant

eh-RID-uh-nuss
Eri
Eridani

Evening Season•December–March

Sky Track•Eridanus follows a southern sky track, rising in the southeast and setting in the southwest.

Lore•The chain of stars that forms Eridanus was known as a river throughout the ancient Mediterranean and Middle Eastern worlds. It was frequently associated with a large river of a particular region such as the Nile, the Euphrates, or the Po. Since one end of Eridanus is close to Rigel, in Orion, the constellation has been called the River of Orion. Another association was with the River Ocean of Homer, believed to flow around the ancient Mediterranean world.

Description• Eridanus is one of the longest and faintest of the constellations. Its southern terminus is, however, a very bright star, Achernar (magnitude 0.5), whose declination is − 57 degrees. This is nearly as far south as the Southern Cross. Neither of these features rises above the horizon for observers in most of the continental United States. They are only visible from the southernmost parts of the country.

From third-magnitude Cursa, near Orion, the stars of Eridanus extend towards the west, then meander south and east, finally disappearing below the horizon.

Betelgeuse   December 8

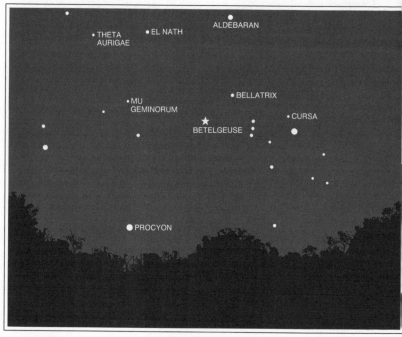

East   **East-Southeast**   **Southeast**

Betelgeuse	0.8 magnitude	−5.7 absolute
BET-el-jooz	yellow-orange M2	magnitude
Alpha, α	supergiant Ia-Iab	05h 55m +07° 24′
Orionis, page 332	650 light-years	February 4 transit
	+0.025 +0.010 +21	STR 305

Location•This first-magnitude supergiant star shines with a yellow-orange sparkle at the right shoulder of Orion, 7 degrees to the east of Bellatrix.

Meaning•The name Betelgeuse is based on an early Arabian description of this star's position at the Hand of the Central One, Orion. Paul Kunitzsch, a scholar at the University of Munich who has made extensive studies of the etymologies of star names, found that when the Arabic title of this star was transliterated in Paris in 1246, the translator incorrectly used a *B* instead of a *Y* at the beginning of the star's title. In later centuries, scholars searched in vain for the correct etymology of an essentially artificial name. Kunitzsch believes that in about 1600, Joseph Scaliger introduced the current spelling of

Betelgeuse to reflect what he incorrectly believed to be the star's original Arabic description, the Armpit of Orion.

Lore•Betelgeuse and some of its neighboring stars were known in Mesopotamia as the Constellation of the King. In India, Betelgeuse was called the Arm, a title also used by the Persians. In ancient Greece, Betelgeuse marked the Right Shoulder of Orion, and this was the star's designation in Ptolemy's *Syntaxis* of the second century A.D. Betelgeuse and its neighbors form the most captivating array of stars in the entire sky. Their brilliance was noted in 425 B.C., by the poet Polymester:

> *. . . through the ether to the lofty ceiling*
> *Where Orion and Sirius dart from their eyes*
> *The flaming rays of fire.*

History•Betelgeuse was first described to be a variable star by John Herschel in 1836. In 1849 he wrote that the star's changes in magnitude were especially noticeable from 1836 to 1840, after which time they settled down. The variability soon increased again, and by the end of December 1852, Herschel considered Betelgeuse to be the brightest star in the night sky north of the celestial equator. Observations by the American Association of Variable Star Observers indicate that Betelgeuse attained a maximum magnitude of about 0.2 during 1933 and again in 1942.

Description•Betelgeuse is an enormous supergiant, which is, on the average, the seventh-brightest star visible in our night sky. It is a semiregular variable with an average apparent magnitude of 0.8 and a usual range of brightness between magnitudes 0.4 and 1.3 over a period of about 6.4 years. This represents a brightness change of over 200 percent during this time.

The brightness of Betelgeuse varies as the star expands and contracts by plus or minus 20 percent of its average diameter. During the period of its variations, the diameter of Betelgeuse ranges between about 700 and 1,000 times that of the sun. If it were in the solar system, centered at the sun, the surface of Betelgeuse at its maximum size would extend beyond the orbit of Jupiter.

The spectral class of Betelgeuse is M2, which indicates a surface temperature of about 3,400°K. At this temperature nearly 90 percent of the star's energy is emitted as invisible infrared radiation.

Betelgeuse is estimated to have a mass somewhere between 10 and 15 times that of the sun. A star of this size may be expected to have an age in the range of approximately 4 to 10 million years. Supergiants of this type are in the final stages of their evolution.

Mintaka   December 9

East   East-Southeast   Southeast

Mintaka min-TAK-ah Delta, δ Orionis, page 332	2.2 magnitude blue-white B0 giant III 800 light-years $-0.003 -0.001 +16$	-4.4 absolute magnitude 05h 32m $-00°$ 18′ January 29 transit STR 307

Location•Mintaka is the westernmost of the three second-magnitude stars that form the Belt of Orion. It is located 7 degrees to the south of Bellatrix.

Meaning•The name Mintaka means the Belt, and it is taken from the Arabian description of the Belt of the Central One.

Lore•Orion's Belt is one of the best-known asterisms in the sky. It is formed by a line of three stars, nearly equal in brightness, separated from each other by about 1.5 degrees of arc.

Various Arabic names for the stars were the Line, the Golden Grains, the Spangles, the String of Pearls, and the Scale-Beam. In China, Orion's Belt was called the Weighing Beam, while on the subcontinent of India, it was known as the Three-Jointed Arrow.

History•The presence of interstellar gas in the Milky Way was discovered as a result of observations of the spectrum of Mintaka, carried out by the German astronomer Johannes F. Hartmann at the Potsdam Observatory during 1904.

Mintaka is a spectroscopic binary, and the absorption lines of these stars appear to double or at least broaden at regular intervals due to Doppler shifts caused as one component star in the system moves towards us and its companion moves away. When the stars are moving across our line of sight as they orbit their center of mass, there are no Doppler shifts and their spectral absorption lines merge.

In the case of Mintaka, Hartmann found that a certain calcium absorption line did not split as did other lines in the spectrum. He reasoned that the stationary line was produced by the absorption effect of calcium atoms in a very thin gas distributed in space along our line of sight to Mintaka.

Although this calcium absorption is relatively easy to detect, it is a minor constituent of interstellar gas. The mixture of gases found in space, between stars in the Milky Way, is revealed by radio telescopes to consist mainly of hydrogen.

Hartmann's discovery was the first evidence for the existence of large quantities of gas in interstellar space.

Description•Mintaka is one of the luminous and massive young blue-white stars belonging to the Orion OB1 Association. It is a giant estimated to have a surface temperature of 28,000°K and a luminosity about 5,000 times the intrinsic brightness of our sun.

This star has an apparent magnitude of 2.2 and a distance from earth of approximately 800 light-years. Mintaka is a spectroscopic binary, and its secondary component is a blue-white main-sequence star. A seventh-magnitude blue-white main-sequence visual companion lies at a distance of 52 arcseconds from Mintaka.

Alnilam December 10

East	East-Southeast	Southeast
Alnilam al-NIGH-lam Epsilon, ε Orionis, page 332	1.7 magnitude blue-white B0 supergiant Ia 1,600 light-years −0.003 −0.002 +26	−6.7 absolute magnitude 05h 36m −01° 12′ January 30 transit STR 315

Location•Alnilam is the star at the center of the three that form the Belt of Orion.

Meaning•The name Alnilam originated with an Arabic description of the Belt stars of Orion as a String of Pearls.

Lore•The glorious symmetry of Orion's bright stars centers on Alnilam, which marks the buckle on the giant Hunter's Belt.

Some of the biblical associations that have been made with these stars include Jacob's Staff, Peter's Staff, the Three Magi, and the Three Kings.

In Scandinavian sky traditions, Mintaka, Alnitak, and Alnilam were called Freya's Staff and were associated with the Norse goddess of love. Greenland Eskimo folklore describes the three Belt stars as representing Seal Hunters who became lost at sea and found their way into the heavens. To Australian aborigines, these stars are said to have shown Three Young Men Dancing. The Pleiades were seven young maidens who played the tunes for these celestial dancers. Chinook Indians of the Columbia River region of Oregon portrayed the Belt and Sword of Orion as Two Canoes.

Description·Alnilam is one of the blue-white supergiants of the Orion OB1 Association. These stars are massive stellar beacons of both spectral classes O and B that have recently evolved away from the main sequence. They are concentrated near the Orion complex of molecular clouds out of which they formed less than 10 million years ago.

The apparent magnitude of Alnilam is 1.7, which, at the star's distance of about 1,600 light-years, represents a luminosity 40,000 times greater than the sun's. Alnilam's spectral designation, B0, suggests a surface temperature of approximately 28,000°K.

Supergiants having such temperatures and luminosities are believed to contain sufficient material so that even after a period of mass loss during their red giant stage, enough material will remain to cause a supernova explosion. The explosive destruction of supergiants is believed, in some cases, to trigger the formation of new generations of stars. Shock waves from supernovae may pass through molecular cloud complexes and cause condensation of dust and gas to occur. As these pockets of interstellar material contract, they increase in temperature and pressure, possibly becoming protostars. If the objects contain more than about 10 percent of the sun's mass, further contraction will raise their core temperatures to about 10 million degrees Kelvin and thermonuclear fusion of hydrogen will begin at their centers. As this occurs, the objects become main-sequence stars.

Such a process of star formation is believed to continue until the molecular clouds in a particular region of the Milky Way are either consumed or dissipated.

Alhena December 12

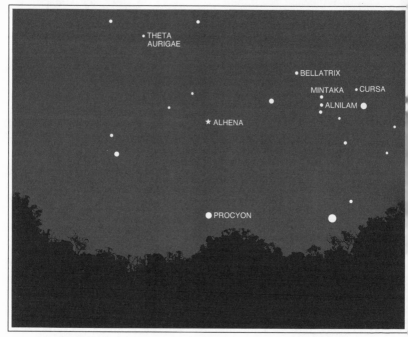

East-Northeast	East	East-Southeast

Alhena al-HEN-ah Gamma, γ Geminorum, page 336	1.9 magnitude blue-white A0 subgiant IV 100 light-years 0.043 − 0.044 − 13	−0.4 absolute magnitude 06h 38m +16° 24′ February 15 transit STR 321

Location•Alhena is located 14 degrees to the northeast of Betelgeuse and is a second-magnitude star.

Meaning•Alhena's name was adapted from a description of this star, and several of its neighbors, as a Brand on the neck of a celestial camel that was portrayed in this part of the sky.

Lore•In Babylon, Alhena and Eta Geminorum were described as the Little Twins, in comparison to Castor and Pollux, the Twins of Gemini. Ptolemy identified Alhena as the star in the right foot of the eastern Twin.

Arabian sky-watchers saw Alhena as part of the Camel's Brand.

Description•Alhena is the third-brightest star in the constellation Gemini and it shines with an apparent magnitude of 1.9. It is a blue-white subgiant whose luminosity is approximately 120 times that of the sun. Alhena's parallax indicates a distance of 100 light-years, and its spectrum suggests a surface temperature of about 9,900°K.

   This star is a spectroscopic binary with an orbital period of 12.4 years.

Nearby Features•The regions of sky to the south and west of Alhena are thought to contain a number of sites that produce new stars. Astronomers using infrared and radio telescopes have, since the 1970s, mapped vast molecular cloud complexes in this area, which provide raw materials for star formation.

   M42, the Orion nebula, is a spectacular example of such a stellar nursery, and is located near the star Iota Orionis in the Sword of the constellation Orion. The following are also major regions of interest as a result of their probable production of new generations of stars.

NGC 2264 AND THE CONE NEBULA (6h 40m +9° 55′)

These features are located about 6 degrees directly to the south of Alhena. NGC 2264 is a young star cluster that contains over one hundred members. The cluster's brightest star, known as S Monocerotis, is a massive and very luminous subgiant with an apparent magnitude of 4.6.

   In addition to the presence of massive type-B main-sequence stars, another clue to NGC 2264's young age is the fact that most of its stars are evidently still undergoing gravitational contraction and have not yet reached the main-sequence stage. As they reach the main sequence, contraction will cease and the thermonuclear fusion of hydrogen will provide all of the star's energy.

   An indication of pre-main-sequence stars in NGC 2264 is provided by observations of many variable stars of the T Tauri variety. These stars are emerging to join the cluster after a process of contraction from interstellar matter.

   NGC 2264 is at a distance of approximately 2,500 light-years from earth, and it is surrounded by quantities of interstellar dust that shine with the reflected light from cluster stars. The cluster's apparent diameter is about 30 minutes of arc, about the same as that of the moon.

   The Cone Nebula is a striking dark feature located at the southern edge of NGC 2264. It is believed to be a type of shadow in the interstellar medium caused by pressure from stellar winds or shock waves. It may represent a region of condensation where additional new stars may form.

   See the Gemini constellation map, page 336.

Alnitak   December 13

| East | East-Southeast | Southeast |

Alnitak	1.8/2.1 magnitude	−6.4 absolute
al-nih-TAK	blue-white O9.5	magnitude
Zeta, ζ	supergiant Ib	05h 41m −01° 57′
Orionis, page 332	1,500 light-years	January 31 transit
	−0.001 −0.002 +18	STR 324

Location•Alnitak is the southeastern star of the three in the Belt of Orion.

Meaning•Giuseppi Piazzi conceived the name Alnitak for the star Zeta Orionis at the beginning of the nineteenth century. He based the name on a transliteration of the Arabic word for belt. An alternate version, Mintaka, was used by Piazzi to identify Delta Orionis.

Lore•It is said that sailors once referred to the Belt of Orion as the Golden Yardarm. In France, the three stars were identified as a celestial Rake, and in parts of Germany the title Three Mowers was used.

An asterism known in Israel as the Arrow in Orion includes the Belt stars and Eta Orionis as the arrow head and Orion's Sword as the arrow's shaft. Tennyson, a devoted watcher of the skies, wrote these lines about the Belt stars of Orion:

> *Those three stars of the airy Giant's zone*
> *That glitter burnished by the frosty dark.*

History•During the early nineteenth century, it was discovered that Alnitak is a close triple star system.

Description•The primary star in the Alnitak system is a blue-white supergiant star of spectral type O9.5, with an apparent magnitude of 1.8. This great star is 30,000 times more luminous than the sun, and it lies at a distance of about 1,500 light-years from earth. The star's spectrum indicates a surface temperature of 29,500°K.

An analysis of the orbital characteristics of this binary system results in an estimated mass 18 times that of the sun for the Alnitak primary star. It is a member of the Orion OB1 star association, and its copious outpourings of ultraviolet radiation have ionized clouds of hydrogen in its vicinity. As a result, these nebulae glow with a soft visible light at temperatures of about 10,000°K.

The secondary star in Alnitak has an apparent magnitude of 4.2, and it is a blue-white giant at a distance of 2.4 arcseconds from the primary. Its mass is estimated to be 6 times that of the sun. This star is also a member of the Orion OB1 Association.

The third component near Alnitak is a tenth-magnitude star at a distance of 58 arcseconds from the primary. Lack of observational evidence for orbital motion indicates that this star is probably an optical companion, with no gravitational connection to the Alnitak system.

Nearby Features•Alnitak is surrounded by a lovely region of nebulosity, visible on photographs because of a reflection of the star's light by dust particles and also from the fluorescence of hydrogen gas by ultraviolet light from this star.

THE HORSEHEAD NEBULA (5h 42m −2° 30′)

A wonderfully shaped figure in the form of a horse's head is located less than 1 degree to the south of Alnitak. This is the Horsehead Nebula, seen on photographs as a jet of unilluminated dust that, at a distance of about 1,200 light-years, partly blocks our view of a portion of bright nebula near Alnitak.

The Horsehead Nebula was discovered on photographs made around the turn of the century.

Rigel

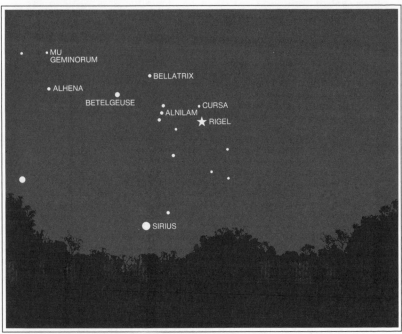

East-Southeast	Southeast	South-Southeast
Rigel RYE-jel Beta, β Orionis, page 332	0.1 magnitude blue-white B8 supergiant Ia 800 light-years $-0.003 -0.002 +21$	-7.0 absolute magnitude 05h 15m $-08°$ 12′ January 25 transit STR 325

Location•Rigel is the fifth-brightest star in our nighttime sky and it shines with a blue-white brilliance 9 degrees to the southwest of Orion's Belt.

Meaning•The name Rigel is taken from the Arabian words for the Left Foot of Orion. Paul Kunitzsch notes that it is one of the few star names that has been in continual use with nearly the same spelling since the tenth century.

Lore•In Scandinavian celestial mythology, Rigel represented the big toe of the giant Orwandil. His other big toe was broken off by the god Thor after it became frostbitten. The thunder god is said to have thrown the toe into the northern sky where it became the star Alcor, companion to Mizar in the Big Dipper.

Description•Rigel is one of the most luminous stars known, and its intrinsic brightness is about 55,000 times greater than that of the sun. It is a blue-white supergiant with an apparent magnitude of 0.1. If Rigel appeared in our sky at the distance of Sirius, it would shine 150 times more brilliantly than the planet Venus. The actual distance of Rigel is estimated to be 800 light-years.

The surface temperature of Rigel is about 10,100°K, and shifts in radial velocity with a period of about 10 days suggest that pulsations in the star's surface occur during this interval of time.

A seventh-magnitude blue-white double star is located at an apparent separation of 9.5 arcseconds. These stars have the same proper motion as Rigel and are probably binary companions with a long orbital period. The overpowering brightness of Rigel makes it difficult to see these companions, which were first identified as double by S. W. Burnham in 1871, using a 6-inch refracting telescope.

Rigel is a member of the Orion OB1 Association, and it is a product of star formation in the molecular cloud complexes of Orion.

Nearby Feature

GEMINID METEORS

The Geminid Meteor shower reaches its peak intensity around the night of December 14, and it is usually one of the best meteor displays of the year.

Geminid meteors are seen to radiate across the sky from a point of origin near the star Castor in the constellation Gemini. It has been suggested that an asteroid known as 1566 Icarus may be the worn-out nucleus of an extinct comet that once spawned this set of meteors.

Observers may expect to see about fifty meteors an hour from this shower, plunging through the atmosphere at speeds averaging 35 kilometers per second.

Iota Orionis   December 16

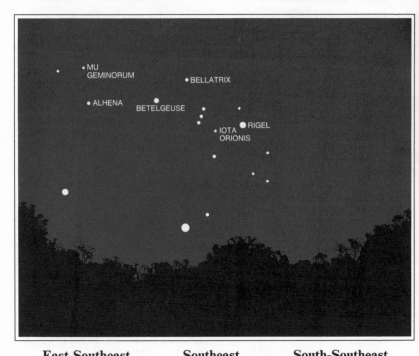

East-Southeast	Southeast	South-Southeast

Iota Orionis eye-OH-tah oar-ee- OH-nis Iota, ι Orionis, page 332	2.8 magnitude blue-white O9 giant III 1,700 light-years −0.004 +0.001 +22	1.0 absolute magnitude 05h 35m −05° 55′ January 30 transit STR 336

Location•Iota Orionis is a third-magnitude star located in the Sword region of Orion, about 5 degrees to the south of Alnilam.

Lore•Iota and neighboring objects in its immediate vicinity were known to Greek astronomers as the Sword of Orion. This description was adopted in Arabia, where Iota Orionis was called the Bright One of the Sword.

History•Wilhelm von Struve, a German astronomer who directed the Pulkowa Observatory near St. Petersburg in Russia, found in 1831 that Iota is a triple star.

Description•The three components of Iota Orionis have shown no change in relative position since they were observed by Struve. The primary has an apparent magnitude of 2.8, and it is a blue-white giant about 15,000 times more luminous than the sun. Stars such as this have surface temperatures estimated to be 31,000°K. Iota Orionis is a member of the group of massive young stars that comprise the Orion OB1 Association. The comparison of Iota's apparent magnitude with the absolute magnitude characteristic of this type of star suggests a distance from earth of approximately 1,700 light-years.

The companion star, which is known as Iota Orionis B, is a blue-white subgiant of magnitude 7.3, and it has a separation of 11.4 arcseconds from the primary.

This AB pair is an attractive sight for observers using telescopes of even a few inches in diameter. The primary, Iota Orionis A, is a spectroscopic binary with an orbital period of 29.14 days.

The third star is of magnitude 11 and has a separation of about 50 arcseconds from the primary.

Nearby Features•Iota Orionis is located on the fringe of the Great Nebula in Orion, and it is one of the extremely hot young stars whose ultraviolet radiation stimulates hydrogen gas in the nebula to glow.

M42: THE GREAT NEBULA IN ORION (5h 34m − 5° 24′)

The Great Nebula in Orion is considered by most telescopic observers to be the loveliest object in the sky. With the naked eye, the nebula resembles a hazy star, and surprisingly, it evidently was not mentioned before the telescopic era, even by astronomers such as Hipparchus, Ulug-Beg, and Tycho Brahe. (See the Orion constellation map, page 332).

Nicholas Peiresc is credited with the discovery of the Great Nebula in Orion in the year 1611. By 1656, Christian Huygens of the Netherlands described the nebula and the remarkable quadruple star, Theta Orionis, seen at its center. Charles Messier's catalogue of nebulae and star clusters contained a description of "the beautiful nebula of the Sword of Orion." Its position was measured by Messier in 1769, and the nebula was assigned numbers 42 and 43 in his catalogue. Across the Channel, William Herschel made the Orion nebula his first recorded observation while using a telescope of his own construction, and in later years Herschel often tested his new telescopes by examining this nebula.

In 1825, John Herschel described the Orion nebula with these words:

> I do not know how to describe it better than by comparing it to a curdling liquid, or to the breaking up of a mackerel sky when the clouds of which it consists begin to assume a cirrus appearance . . .

Tennyson wrote the following about the Belt, Sword, and Nebula of Orion:

> . . . *A single misty star*
> *Which is the second in a line of stars*
> *That seem a sword beneath a belt of three,*
> *I never gazed upon it but I dreamt*
> *Of some vast charm concluded in that star*
> *To make fame nothing . . .*

The introduction of photography and spectroscopy to astronomy during the latter part of the nineteenth century revealed marvelous details of the Orion Nebula beyond the province of even the sharpest-eyed observer.

On a September night in 1880, Henry Draper used the 11-inch Clark refractor at his Hudson Valley observatory to make a 51-minute photographic exposure of the Great Nebula, and by 1883, Andrew A. Common in England was able to photograph details of the nebula invisible to the naked eye.

About this same time, William Huggins succeeded in photographing bright emission lines in the Orion nebula's spectrum from his observatory near London. This demonstrated that much of the nebula consists of brightly glowing hydrogen and other gases.

Since the 1960s, astronomers have applied new instruments and techniques to explore space in and around this nebula. Methods of infrared and radio astronomy have been found to be especially useful in this research.

As a result of recent studies, it has been shown that the Great Nebula in Orion sits like a hot button on the front surface of a vast complex of molecular clouds that stretches across much of the constellation Orion. The Great Nebula lies at a distance of approximately 1,600 light-years from earth, and it has a diameter of about 30 light-years. Hidden from our view, just behind the visible portions of the nebula, is a region of new star production of much interest to astronomers.

This site contains stars estimated to be less than 100,000 years old, stellar babies in the cosmic time scale. These stars are so young that they have not had sufficient time to blow away the dust that hides them from our view. As visible light from stars in this young cluster strikes the dust, some of this energy is converted to heat, which may then be detected from earth with the use of infrared telescopes.

In 1966, a point source of infrared energy with a temperature of about 600°K was discovered in the Orion nebula. Known as the Becklin-Neugebauer object after its discoverers, Eric E. Becklin and Gerald Neugebauer of Cal Tech, it appears to be a collapsing protostar, and similar objects have since been found in the area.

The Great Nebula in Orion is a region where star production probably began less than one million years ago. The stars of the Trapezium system, which illuminate the Orion nebula, are believed to have been formed during this episode.

Ultraviolet energy from these hot, young Trapezium stars may now be contributing to a second generation of star birth, currently taking place behind the Orion nebula, in the adjacent molecular cloud. Observations made with the Kuiper Airborne Observatory revealed the infrared cluster in this cloud, hidden from the view of optical telescopes by interstellar dust.

Trapezium stars are apparently generating a surrounding region of hot expanding hydrogen called a Strömgren sphere, which, as it pushes into the molecular cloud, generates shock waves that may have triggered the formation of stars in the young infrared cluster.

In 1946, Bart J. Bok of Steward Observatory and the University of Arizona first called attention to a number of round dark nebulae that were first discovered by E. E. Barnard. Bok suggested that these objects are regions of condensation, leading to the formation of protostars. The nebulae, which are often called Bok globules, vary in diameter, the smallest being about one-half light-year across and the largest examples having diameters of several light-years. Bok globules differ from other dark objects seen in certain nebulae, such as the Rosette in Monocerous. The Rosette Nebula's spots are believed by some astronomers to be "elephant trunk" structures as seen from the end.

Yet another possible cause of star formation in some parts of the sky may be collisions between giant molecular clouds. As these clouds pass and encounter each other, gas and dust at their contact surfaces may be compressed in a way that triggers protostar condensation.

THETA ORIONIS: THE TRAPEZIUM (5h 35m −5° 23′)

Theta Orionis is a multiple star system located in the heart of the Orion nebula. It consists of four stars with a magnitude range of 5.1 to 6.7, with the average of their separations being about 14 arcseconds. These four stars are visible to observers using small telescopes and, due to the geometric pattern they form, are known as the Trapezium.

The entire star system of Theta Orionis is believed by astronomers to contain a total of at least twelve components, all young stars formed from gas and dust in the region of the Orion nebula. It is the ultraviolet energy from the most luminous of these stars that causes the Orion nebula to glow.

Castor   December 17

Northeast   **East-Northeast**   **East**

Castor	1.6/2.6 magnitude	1.9 absolute
CASS-ter	blue-white A2	magnitude
Alpha, α	main sequence V	07h 35m +31° 53′
Geminorum, page	46 light-years	March 1 transit
336	0.170 −0.102 −1	STR 337

Location•Castor is one of the twin stars of Gemini, along with Pollux. It is a second-magnitude star located 21 degrees to the northeast of Alhena.

Meaning•The name Castor was widely used in reference to this star in ancient Greece and Rome.

Lore•Castor was known in Mesopotamia as the Western One of the Twins, and it was described by Ptolemy as the star in the head of the western Twin. Arabian astronomers called Castor the Head of the Foremost Twin.

In classical mythology, Castor was the Gemini brother who was an expert trainer of horses. He was the mortal Twin, son of King Tyndareus of Sparta and his Queen Leda.

The name Apollo was also used for this star until the mid-eighteenth century.

History•Castor is a multiple star that was first resolved by the Italian astronomer Giovanni Domenico Cassini in 1678, while he was director of the Paris Observatory. Castor's binary nature was evidently not studied again until 1718, when James Bradley began observations in which he recorded the component's positions over a period of years.

William Herschel also made extensive observations of Castor, and by 1803 he was convinced that the star's components moved in gravitationally bound, elliptical orbits around their common center of mass. This was the first observational evidence that the force of gravity acts in distant space in the same way as it does within the solar system.

Description•There are three visible components in the Castor system and each of these is a spectroscopic binary. These pairs of stars are designated Castor A, B, and C. Stars A and B form the set discovered by Cassini. Component C has an apparent magnitude of 8.8 and is at a distance of 73 arcseconds from the primary.

Component A has a combined apparent magnitude of 1.9 and consists of two close blue-white main-sequence stars. With a magnitude of 2.9, component B also consists of two blue-white stars that form a spectroscopic binary.

For centuries, the separation between Castor's brightest components could be resolved with a small telescope. However, by 1949 these stars reached a point in their orbits where the apparent distance between them became less than 3 arcseconds; their minimum separation occurred in 1969. During this time the components of Castor became a difficult subject for small instruments. The separation is now increasing, and after about 1990, Castor again will become a favorite double star for observers using telescopes of less than 8 inches in diameter.

Castor C is also a spectroscopic binary whose orbital plane is nearly parallel to our line of sight, and as a result the components eclipse each other as they revolve in their orbits. These stars are both yellow-orange main-sequence dwarfs, with an orbital period of 19 hours and 32.6 minutes. Their orbit's orientation lends itself to precise measurement, and astronomers have determined that the components have masses about 60 percent that of the sun. These relatively small stars have luminosities equal to only about 3 percent of the sun's, and their diameters are approximately 70 percent of our sun's size.

The blended light of the components of Castor C appear as a single star, which is visible in telescopes of 3 inches in diameter or larger.

Pollux

Northeast	East-Northeast	East

Pollux	1.2 magnitude	1.0 absolute
POL-lucks	yellow K0	magnitude
Beta, β	giant III	07h 45m +28° 02′
Geminorum, page	35 light-years	March 4 transit
336	−0.627 −0.051 +3	STR 357

Location· Pollux is the brighter of the Twin stars of Gemini, and it is located 4 degrees to the southeast of Castor.

Meaning· The name Pollux is that of the immortal Twin of the Gemini, who was said to have been a son of Zeus and Leda of Sparta. He was described as the brother of Helen of Troy and the half-brother of Castor.

Lore· The Twin stars in Gemini form such a striking pair that they have been associated with each other throughout history. In Babylon, Pollux was called the Eastern One of the Twins and also the Yoke of the Enclosure. Ptolemy described Pollux in his *Syntaxis* as the "red star in the head of the eastern Twin." This probably was an exaggeration of the star's yellow color. In Arabia, Pollux was the Head of the Hindmost Twin. Pollux was characterized as being hindmost because it transits the meridian about 10 minutes after Castor.

Sometimes this star was called Hercules, the companion of Apollo, an alternate name for Castor in Greek astronomical lore.

Description· Pollux is the twelfth-brightest star in our night sky, and with an apparent magnitude of 1.2, it is nearly 50 percent brighter than its neighbor Castor.

This star is a yellow giant with a surface temperature of approximately 4,500°K, and it is about 35 times more luminous than the sun.

The trigonometric parallax of Pollux indicates a distance of 11 parsecs, or 35 light-years. This distance is close to the standard used for the calculation of absolute magnitudes, 10 parsecs or 32.6 light years. As a result, the apparent and absolute magnitudes of Pollux have nearly the same value. Pollux also happens to be the closest giant star to the earth.

There are several faint stars apparently close to Pollux in a telescope's field of view. Studies of their space motions indicate, however, that none of these stars is physically related to Pollux.

December 21 is the day when the sun usually reaches its most southerly position on the ecliptic and winter begins in the northern hemisphere. At the moment when our winter season begins, the sun is located at the 18-hour circle of right ascension and has a declination of −23.5 degrees on the celestial sphere. Our nearest star is then in a direction about 1 degree to the northwest of M8, the Lagoon Nebula, above the Teapot's Spout in the constellation Sagittarius.

Although average daily temperatures will continue to drop for about one month due to heat absorption by ground and water, it is cheering to remember that at the instant winter begins, the sun ceases its apparent southerly motion and gradually begins to move north along the ecliptic. This causes the days to lengthen and spring to return eventually to this part of the world.

Saiph

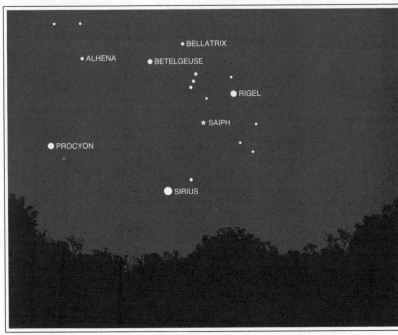

East-Southeast	Southeast	South-Southeast

Saiph	2.1 magnitude	− 6.8 absolute
SAFE	blue-white B0.5	magnitude
Kappa, κ	supergiant Ia	05h 48m − 09° 40′
Orionis, page 332	2,000 light-years	February 2 transit
	− 0.003 − 0.005 + 21	STR 369

Location•Saiph is the last of the bright stars of Orion to ascend into the sky. It is a second-magnitude object located at the southeast corner of the constellation figure, 5 degrees from Iota Orionis.

Meaning•The name Saiph was applied to this star by Piazzi after having been borrowed by him from the translation of the Arabian description for the Sword of Orion.

Lore•In antiquity, this star marked the right knee of Orion. The portrayal of the constellation figure shown in Bayer's *Uranometria* of the early seventeenth century shows Orion facing away from us and the star at the mighty Hunter's left knee.

The rise of Saiph completes the hourglass outline of Orion's magnificent figure. This glorious constellation will rise higher to dominate the winter evening sky, just as its features captivate the imaginations and intellects of sky-watchers.

Description•Saiph is one of the supergiant stars of the Orion OB1 Association. Its luminosity is about 45,000 times greater than that of the sun. This high intrinsic brightness enables us to see Saiph shine with an apparent magnitude of 2.1, even at the star's estimated distance of 2,000 light years. Saiph is a blue-white star with a surface temperature of approximately 27,000°K.

This star marks the southeastern corner of a lobe in the Orion Molecular Cloud Complex; however, Saiph lies several thousand light-years beyond the limits of these clouds. Studies of its radial velocity show that Saiph is moving away from us at the rate of 21 kilometers per second, having been ejected in that direction during an early stage in its evolution within the vicinity of the cloud complex.

In recent years, advances in areas of astronomy, such as photometric classification of stars, astrometry, radio, infrared, and X-ray observations, and theoretical model building have produced a sophisticated picture of the processes of stellar evolution. Many astronomers and astrophysicists have made contributions to these efforts, and it is now possible to trace the general evolution of stars from their birth in clouds of gas and dust, through the main-sequence stage, and then during the transformations that occur as the star becomes a giant, before it expires either by contracting to become a white dwarf or by exploding cataclysmically as a supernova.

Astronomers are our representatives in the exploration of the universe. Their intellects, instruments, and imaginations enable them to visit the distant shores of wondrous worlds.

As we gaze at stars in our night sky, and dream of their nature and meaning, we also take part in this grand endeavor.

Part III
Appendixes

Appendix A

Star Time Reference (STR) Numbers

After you have become familiar with the primary identification system of *The Star Guide,* you may wish to make use of the STR numbers provided with each featured star's description in order to write an observing schedule to help you to locate a number of bright stars at various times during a single night.

The STR numbers found in *The Star Guide* indicate the separation in minutes of time between the Sky Screen arrival of featured stars. For example, the STR number of the star Arcturus is 771 and that of Spica is 832. Subtraction shows that there is a difference of 61 between these numbers, which indicates that on any night during the year, Spica will reach its Sky Screen map position 61 minutes after Arcturus attains its corresponding location in the eastern sky.

Use of the STR numbers enables you to write an observing schedule for one night that will tell when to use the Sky Screen maps and position descriptions for a selected set of stars. In this way, identification clues of a featured star may be put to use during a period of many months and not just on the single date suggested in the primary identification system of *The Star Guide.*

Example: Suppose that it is the night of April 5–6 and you would like to spend several hours identifying stars with the help of the maps and descriptions of *The Star Guide.*

1. Turn to the Time-Date charts (next page) and place a straightedge parallel to the line that marks the start of April, and then move it to the right about halfway between that line and the set of marks indicating the night of April 10–11.

2. Notice that the "time line" of the star Arcturus intersects your straightedge at 9 P.M. for the night of April 5, the date when this star is featured in the basic *Star Guide* system of identification.

3. Also note that time lines of other stars also intersect the straightedge at locations corresponding to various times throughout the night of April 5–6. For example, Spica's line intersects at about 10 P.M., and that of Vega meets the straightedge near 12:30.

4. Prepare a list showing when the star lines meet the line of April 5–6.

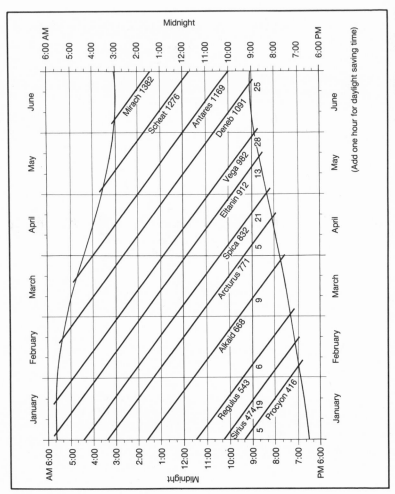

Time-Date Chart, January-June

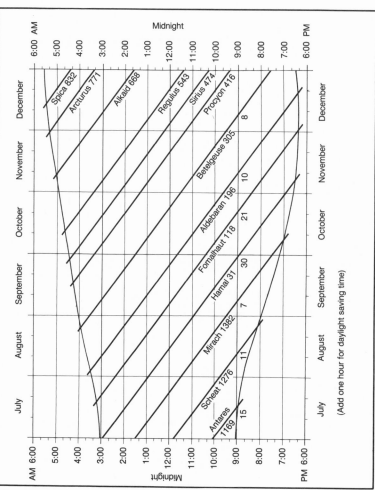

Time-Date Chart, July-December

Arcturus	9:00 P.M.
Spica	10:00
Eltanin	11:30
Vega	12:30 A.M.
Deneb	2:20
Antares	3:40

Note that beginning and ending of astronomical twilight at mid-northern latitudes throughout the year are indicated by the curved lines at the top and bottom of the time-date charts.

The list given above is a schedule of Sky Screen arrival times for a selection of stars featured in *The Star Guide.* It indicates when on the night of April 5–6 the Sky Screen maps and descriptions of star positions are valid. (For reasons of clarity, the charts show only a fraction of the stars featured in this book.)

Making an observing schedule such as that given above enables you to further identify or "bracket" additional stars. Suppose you wish to use *The Star Guide* from about 9 to 10 P.M. on the night of April 5. From the list you make with the help of these charts, it may be seen that stars in *The Star Guide* between the descriptions of Arcturus and Spica arrive at their designated Sky Screen locations on the night of April 5 between the hours of 9 and 10 P.M.

A more detailed observing schedule may now be constructed by listing these stars along with their STR numbers, and using the STR numbers to determine the various Sky Screen arrival times of these stars on the night of April 5.

Table A-1 gives an example of such a chart for an observer in Denver, where the basic Sky Screen time is 9 P.M.

Table A-1

Observer's location: **Denver, Colorado**
Basic Sky Screen time: **9:00 P.M.**
Date: **April 5**

Star	STR number	Time difference Minutes	Time difference Hours/minutes	Sky Screen arrival time
Arcturus	771	0	0/00	9:00
Izar	779	8	0/08	9:08
Eta Draconis	785	14	0/14	9:14
Giena	814	43	0/43	9:43
Alphecca	829	58	0/58	9:58
Spica	832	61	1/01	10:01

1. Write the names and STR numbers of stars that arrive on the Sky Screen from 9 to 10 P.M. on April 5 (columns 1 and 2).

2. Subtract the STR number of Arcturus (771) from that of subsequent stars to indicate the number of minutes separating these stars' arrival times from that of Arcturus.

3. Add these time differences to the time when you plan to begin your observations (9 P.M. in this example).

What to Do If Subtraction of STR Numbers Gives a Negative Result

The STR system is based on the number of minutes in 24 hours, 1,440. Its purpose is to provide a numerical frame to help calculate relative arrival times of stars at their designated Sky Screen positions.

At times, a subtraction of STR numbers performed in setting up an observing schedule as shown above will result in a negative number. For example, on the night of September 17, the featured star Almach has the STR number 1422. Subtracting this from the STR number of Sheratan, 25, gives an answer of $25 - 1422 = -1397$. Whenever a negative result occurs when you are subtracting STR numbers, simply add 1440 to the result. This provides a positive value for the number of minutes separating the Sky Screen arrival time of the two stars. In our example we add: $-1397 + 1440 = 43$. This tells us that Sheratan will reach its Sky Screen position 43 minutes after Almach has moved into its designated Sky Screen position.

In Table A-2, an observer in Boston on the night of September 17 may write this observing schedule:

Table A-2

Observer's location:	Boston, Massachusetts
Basic Sky Screen time:	9:00 P.M.
(after adding one hour for daylight saving time)	10:00 P.M.
Date:	September 17

Star	STR number	Time difference Minutes	Hours/ minutes	Sky Screen arrival time (DST)
Almach	1422	0	0/00	10:00
Sheratan	25 + 1440 = 1465	43	0/43	10:43
Hamal	31 + 1440 = 1471	49	0/49	10:49
Mirfak	41 + 1440 = 1481	59	0/59	10:59
Algol	48 + 1440 = 1488	66	1/06	11:06

Use of Time-Date Charts and STR Numbers to Help Find Stars on the Sky Screen before Your Basic Observing Time

During the long nights of autumn and winter, you can use the charts and STR numbers to locate stars early in the evening hours. The STR numbers will help you to set up an observing schedule that takes advantage of the extra hours of darkness.

In Table A-3, such a schedule is provided for an observer in Minneapolis on the night of December 24. Note that in this example, unlike the previous ones, stars are listed in order of decreasing STR numbers. In making the list, read forward in *The Star Guide* from the star that is featured for your date of observation. For example, the star Saiph is featured for the night of December 24, and the star that precedes it is Pollux, which is listed for December 21. Next on your list would be Castor, associated with December 16, and so on, earlier into the stars featured during December, and, if you wish, November.

Your early-evening observing list may be extended in this way until the Sky Screen arrival times that you will calculate occur when evening twilight is still too bright to locate stars.

Table A-3

Observer's location:	Minneapolis, Minnesota
Basic Sky Screen time:	9:00 P.M.
Date:	December 24

Star	STR number	Time difference		Sky Screen arrival time
		Minutes	Hours/minutes	
Saiph	369	0	0/00	9:00
Pollux	357	−12	−0/12	8:48
Castor	337	−32	−0/32	8:28

You may wish to extend a list of this kind to include stars that arrive at their featured positions of the Sky Screen soon after the fading of evening twilight on the date of your observations. The end of twilight is indicated on the time-date charts.

Appendix B

The Brightest Stars Seen from Midnorthern Latitudes

Apparent Magnitude and Relative Brightness

us	−1.45	100%
turus	−0.06	28
a	0.04	25
ella	0.08	24
el	0.11	24
cyon	0.35	19
ir	0.77	13
elgeuse	0.80	13
ebaran	0.85	12
a	0.96	11

Distance from Earth (in light-years)

Sirius	8.8
Procyon	11.4
Altair	16
Vega	26
Arcturus	36
Capella	41
Aldebaran	70
Spica	650
Betelgeuse	650
Rigel	800

Absolute Magnitude and Luminosity Compared to Sun

el	−7.0	52,500×
elgeuse	−6.0	21,000
ca	−3.4	1,900
ebaran	−0.7	160
turus	−0.2	100
ella	0.1	75
nponents)	0.6	47
ga	0.5	50
us	1.4	25
air	2.3	10
cyon	2.7	7

Surface Temperature (K) and Color

Spica	25,500°	blue-white
Rigel	10,100°	blue-white
Vega	9,900°	blue-white
Sirius	9,500°	blue-white
Altair	8,000°	blue-white
Procyon	6,600°	white
Capella	5,600°	white
(companion)	5,000°	yellow-white
Arcturus	4,200°	yellow
Aldebaran	3,800°	yellow
Betelgeuse	3,400°	yellow-orange

The Ten Brightest Stars of Winter

Apparent Magnitude and Relative Brightness			*Distance from Ea~ (in light-years)*
Sirius	−1.45	100%	Sirius
Capella	0.08	24	Procyon
Rigel	0.11	24	Pollux
Procyon	0.35	19	Capella
Betelgeuse	0.80	13	Castor
Aldebaran	0.85	12	Aldebaran
Pollux	1.15	9	Bellatrix
Adhara	1.50	7	Adhara
Castor	1.58	6	Betelgeuse
Bellatrix	1.63	6	Rigel

Absolute Magnitude and Luminosity Compared to Sun			*Surface Temperature (K) and Color*		
Rigel	−7.0	55,000×	Adhara	23,000°	blue-whi
Betelgeuse	−6.0	21,000	Bellatrix	23,000°	blue-whi
Adhara	−5.0	8,300	Rigel	10,100°	blue-whi
Bellatrix	−3.3	1,750	Castor	9,500°	blue-whi
Aldebaran	−0.7	160	Sirius	9,500°	blue-whi
Capella	0.1	75	Procyon	6,600°	white
(components)	0.6	47	Capella	5,600°	white
Castor	0.9	40	(companion)	5,000°	yellow-w
Pollux	1.0	35	Pollux	4,500°	yellow
Sirius	1.4	25	Aldebaran	3,800°	yellow
Procyon	2.7	7	Betelgeuse	3,400°	yellow-o

The Ten Brightest Stars of Spring

Apparent Magnitude and Relative Brightness			*Distance from E◌ (in light-years)*
Arcturus	−0.06	100%	Arcturus
Spica	0.96	40	Alioth
Regulus	1.35	27	Regulus
Alioth	1.78	18	Mizar
Dubhe	1.79	18	Alphard
Alkaid	1.86	17	Kochab
Alphard	1.99	16	Dubhe
Kochab	2.07	14	Algieba
Mizar	2.09	14	Alkaid
Algieba	2.10	14	Spica

	Absolute Magnitude and Luminosity Compared to Sun			*Surface Temperature (K) and Color*	
	−3.4	1,900×	Spica	25,500°	blue-white
l	−1.6	360	Alkaid	20,000°	blue-white
ıs	−0.7	160	Regulus	12,500°	blue-white
a	−0.6	145	Alioth	9,900°	blue-white
ɔ	−0.5	130	Mizar	9,200°	blue-white
rd	−0.5	130	Algieba	4,600°	yellow
	−0.4	120	Dubhe	4,600°	yellow
us	−0.2	100	Arcturus	4,200°	yellow
	−0.2	100	Alphard	4,000°	yellow
	0.0	85	Kochab	4,000°	yellow

The Ten Brightest Stars of Summer

	Apparent Magnitude and Relative Brightness			*Distance from Earth (in light-years)*
	0.04	100%	Altair	16
	0.77	51	Vega	26
es	1.00	41	Rasalhague	60
ɔ	1.25	33	Kaus Australis	150
a	1.62	23	Nunki	250
Australis	1.83	19	Shaula	350
ıs	1.87	19	Antares	400
hague	2.07	15	Sargas	500
i	2.08	15	Sadr	800
	2.23	13	Deneb	1,600

	Absolute Magnitude and Luminosity Compared to Sun			*Surface Temperature (K) and Color*	
ɔ	−7.3	70,000×	Shaula	25,500°	blue-white
res	−4.7	6,300	Nunki	23,000°	blue-white
	−4.7	6,300	Kaus Australis	10,000°	blue-white
ıs	−4.5	5,300	Vega	9,900°	blue-white
la	−3.4	1,900	Deneb	9,000°	blue-white
ki	−2.5	830	Rasalhague	8,500°	blue-white
Australis	−1.5	330	Altair	8,000°	blue-white
	0.5	50	Sargas	7,000°	blue-white
lhague	0.8	40	Sadr	6,000°	white
r	2.3	10	Antares	3,400°	yellow-orange

The Ten Brightest Stars of Autumn

Apparent Magnitude and Relative Brightness			*Distance from Ea (in light-years,*
Fomalhaut	1.16	100%	Fomalhaut
Mirfak	1.80	55	Caph
Hamal	2.00	46	Diphda
Alpheratz	2.03	45	Hamal
Diphda	2.04	44	Mirach
Mirach	2.06	44	Algol
Almach	2.13	41	Alpheratz
Algol	2.20	38	Schedar
Schedar	2.22	38	Almach
Caph	2.26	36	Mirfak

Absolute Magnitude and Luminosity Compared to Sun			*Surface Temperature (K) and Color*		
Mirfak	−4.3	4,350×	Algol	10,100°	blue-wh
Almach	−2.2	650	Alpheratz	10,000°	blue-wh
Schedar	−1.0	200	Fomalhaut	8,900°	blue-wh
Alpheratz	−0.9	190	Caph	7,000°	blue-wh
Algol	−0.3	110	Mirfak	6,500°	white
Mirach	0.1	75	Schedar	4,500°	yellow
Hamal	0.2	70	Diphda	4,350°	yellow
Diphda	0.7	45	Hamal	4,200°	yellow
Caph	1.5	20	Almach	4,000°	yellow
Fomalhaut	1.9	15	Mirach	3,200°	yellow-o

Appendix C

Examples of the
Hertzsprung-Russell Diagram

The following graphs are examples of Hertzsprung-Russell diagrams. The position of stars on the graphs were determined by each star's intrinsic brightness and color. The vertical scale indicates a range of brightness expressed in terms of both absolute magnitude (i.e. $+15$ to -10) and luminosity compared to the Sun (1/10,000 to 1,000,000). The horizontal scale shows a range of spectral classes from O to M and also an approximate color range from blue-white to yellow-orange, indicating the apparent colors of the stars. Surface temperatures are also shown.

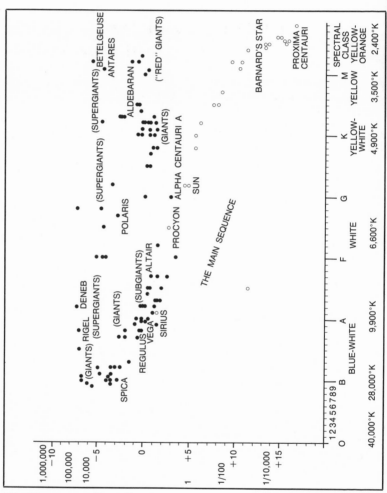

The Hertzsprung-Russell Diagram

Stars on this particular graph are among the brightest and/or closest to the Earth. A number of stars are named, as are regions of the graph associated with several categories of luminosity (i.e. main sequence, sub-giants, giants, and supergiants).

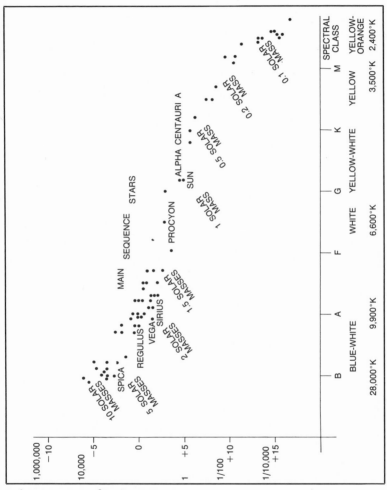

Relative masses of main sequence stars

This version shows a selection of bright and/or nearby stars in the category known as the main sequence. Stars in this set produce energy through the fusion of hydrogen in their cores. This stage occupies most of the lifetimes of all stars. The luminosity of a main sequence star is mostly determined by its mass. The unit of comparison is the mass of the Sun.

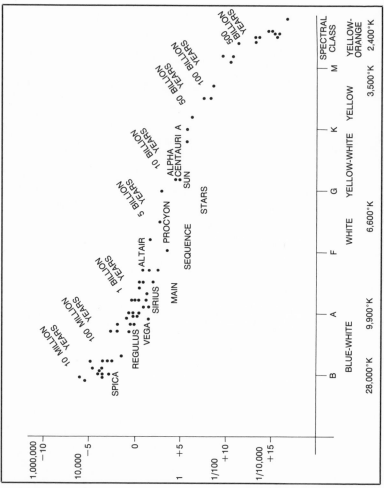

Relative ages of main sequence stars

This H-R diagram show approximate main sequence life expectancies for stars along the main sequence. The estimated age of the universe is between 10 and 20 billion years, so it is evident that the stars at the lower right of the main sequence have never consumed their supplies of core hydrogen and evolved into red giants. On the other hand, as we look toward the upper left, the most luminous stars would be expected to consume their core hydrogen and become red giants in periods of less than a few million years.

Glossary

Aberration of Starlight. A slight variation in a star's apparent position, caused by earth's orbital motion relative to the velocity of the star's light.

Absolute magnitude. The apparent magnitude a star would have if it were seen at a standard distance of 10 parsecs, or 32.6 light-years.

Absolute zero. The zero point of the Kelvin temperature scale, equal to $-273.16°$ Celsius or $-459.69°$ Fahrenheit.

Absorption spectrum. Characteristic dark lines or bands displayed on a continuous spectrum of colors when light is passed through a gas.

Accretion disk. A disk surrounding a star or other celestial object, formed by orbiting particles or gas attracted to the object's vicinity by its gravity.

Angular diameter. The apparent diameter of a celestial object, expressed in degrees, minutes, or seconds of arc (see Degree).

Angular separation. The apparent distance between two stars or other celestial objects, expressed in angular measurements.

Angular size. The apparent size of a celestial object expressed in angular measure.

Aperture. The usable diameter of a telescope's objective lens or mirror.

Apparent magnitude. The brightness of a celestial object expressed in a system in which a difference of one magnitude represents a brightness factor of about 2.5.

Arcsecond. A unit of angular measurement equal to 1/60 of an arcminute or to 1/3,600 of a degree of arc.

Asterism. An eye-catching formation of stars, which may or may not belong to a single constellation.

Astrometric binary. A pair of stars in a binary system that are too close to be separated with a telescope, but whose dual nature is detected by observations of a wavering proper-motion path.

Astrometry. A branch of astronomy that concentrates on measurements of the positions and motions of celestial objects. It is sometimes known as spherical or positional astronomy.

Astrophysics. The branch of astronomy that concentrates on the physical nature of celestial objects.

Atmosphere. An envelope of gas that surrounds a celestial object.

Autumnal equinox. An invisible point on the celestial sphere, located in the constellation Virgo, defined by the intersection of the ecliptic and the celestial equator. The sun's arrival at this point, on or about September 23, marks the beginning of autumn in the northern hemisphere.

Binary star. A double star whose components are gravitationally bound to each other and revolve around a common center of mass.

Black hole. An object whose mass has become so highly concentrated that it theoretically occupies no space at all. The gravitational field surrounding a black hole is believed to be so intense that even light may not escape from a region immediately surrounding the concentrated mass.

Blue shift. A shift of spectral lines towards the shorter-wavelength, blue, end of the spectrum caused by the light source's motion towards the earth.

Celestial equator. The great circle on the celestial sphere that lies half way, or 90 degrees, between the celestial poles.

Celestial mechanics. The branch of astronomy most concerned with the interactions of forces and motions of celestial objects.

Celestial meridian. The great circle on the celestial sphere that passes through both poles, as well as the zenith and north and south points on the horizon of an observer's sky. A celestial object's transit of the celestial meridian represents the object's highest elevation in the sky.

Celestial poles. The two imaginary points on the celestial sphere defined by intersections with extensions of the earth's axis of rotation.

Celestial sphere. An apparent sphere of indefinite size upon which celestial objects appear to be located.

Center of mass (gravity). The point at which all the mass of an object or system of objects may be considered to be concentrated.

Cepheid variable. A type of highly luminous, pulsating supergiant stars, named after their prototype, Delta Cephei. The periods of light variations in these stars are directly related to their intrinsic brightnesses, and, as a result, Cepheids may be used to estimate distances up to about 10 million light-years from the earth. These stars are also known as classical or type I Cepheids.

Chandrasekhar's limit. An upper limit, of 1.4 solar masses, on the theoretical mass of a white dwarf star. If this size is exceeded, a dying star will bypass the white dwarf stage and contract to become a neutron star, or, if it is sufficiently massive, the object will continue to contract to become a black hole.

Charge-coupled device (CCD). An electronic accessory used to increase the sensitivity of a telescope in recording faint visible light and other forms of electromagnetic radiation.

Circumpolar stars. Stars in the vicinity of the celestial pole that are always above the horizon for observers at a particular latitude.

Circumstellar shell. A shell of dust or gas that surrounds a star.

Cluster of galaxies. A group of up to several thousand galaxies, bound together by their mutual gravitational attraction.

Colure. One of two great circles on the celestial sphere that pass through the poles as well as points on the ecliptic that mark the sun's position at the start of each season. The equinoctial colure intersects both the vernal and autumnal equinoxes. The solstitial colure passes through the summer and winter solstices on the ecliptic.

Condensation. A consolidation of matter into a smaller volume.

Configuration. A grouping seen in the sky formed by the temporary juxtaposition of two or more planets, stars, and/or the moon.

Constellation. One of eighty-eight areas in the sky, whose boundaries define regions on the celestial sphere.

Continuous spectrum. A continuous band of colors or other wavelengths of electromagnetic radiation. Continuous spectra are produced by hot solids, liquids, or highly compressed, opaque gases such as constitute the interiors of stars.

Core (of a star). The central region of a star. In main-sequence stars, all energy obtained through the fusion of hydrogen into helium is produced in the stellar cores, where temperatures are at least 10 million degrees Kelvin.

Cosmic background radiation (CBR). This microwave energy is observed coming from all directions in the universe. It was discovered in 1964 and is believed to represent the "stretched out" wavelengths of light from the flash of the primeval fireball that formed our universe.

Cosmic rays. High-energy atomic particles that reach earth from space, traveling at speeds approaching that of light. Some cosmic-ray particles may have originated as a result of supernova explosions.

Cosmology. Study of the origin, structure, and evolutionary processes of the universe.

Dark nebulae. Relatively dense regions of interstellar dust that obscure the light of stars within the nebulae or behind them.

Declination. The coordinate on the celestial sphere which measures angular distances north or south of the celestial equator.

Degree (angular). 1/360 of the circumference of a circle. A degree of arc on the celestial sphere is equal to about twice the apparent diameter of the full moon.

Density. A measure of the concentration of matter.

Density-wave theory. A theory used to explain the spiral structure of galaxies.

Diffuse nebulae. An irregular cloud of gas or dust in a spiral galaxy such as the Milky Way.

Disk (of a celestial object). The apparently circular cross-section of a spherical celestial object.

Disk (of a galaxy). The flattened, circular region containing the majority of stars in a spiral galaxy.

Eclipse. The complete or partial disappearance of a celestial object behind another, or the appearance of the shadow of the earth on the moon.

Eclipsing binary star. A star, such as Algol, that is periodically eclipsed by a companion in a binary system.

Ecliptic. The apparent path of the sun across the celestial sphere.

Electromagnetic radiation. Radiation produced as a result of variations in electric and magnetic fields. The electromagnetic spectrum includes gamma rays, X rays, ultraviolet light, visible light, infrared light, and radio energy. All forms of electromagnetic radiation have a speed in a vacuum of about 3×10^{10} centimeters per second or approximately 186,000 miles per second.

Electromagnetic spectrum. The range of wavelengths manifested by electromagnetic energy.

Electron. A subatomic particle with a negative electric charge, which normally moves in a zone surrounding the nucleus of an atom.

Element (chemical). A substance that cannot be broken down into its components through chemical actions.

Elephant-trunk structures. Long, dark cones that may be associated with star-formation processes. They are seen in some nebulae where H II regions of heated gas are expanding and flowing past comparatively dense regions of cool dust and gas. When seen from the end, elephant-trunk structures may appear as dark spots called globules.

Ellipse. A closed curve representing the orbital paths of planets, satellites, and components of multiple star systems.

Elliptical galaxy. A type of galaxy having an elliptical cross-section and containing very little interstellar gas and dust.

Emission line. A bright spectral line produced at a specific wavelength of electromagnetic radiation.

Emission nebula. A cloud of gas having temperatures of about 10,000°K, whose spectrum is characterized by emission lines.

Emission spectrum. A spectrum that consists of emission lines produced by hot glowing gas of low density.

Equatorial mounting. A telescope mounting having one axis parallel to the earth's axis so that compensation may be more easily made for the earth's rotation.

Event horizon. The surface surrounding a black hole at which the escape velocity is equal to the speed of light.

Extragalactic. Outside of the Milky Way Galaxy.

Eyepiece. The lens or set of lenses in a telescope where you place your eye to look through the instrument.

First Point of Aries. A traditional name for the vernal equinox, the sun's location on the celestial sphere at the start of the spring season in the northern hemisphere. This term originated when the vernal equinox was actually located in the constellation Aries; precession has now shifted this equinox to the constellation Pisces, the Fish.

Force. A push or pull causing a change in an object's speed or direction of travel.

Full moon. The phase of the moon at the point of its orbit at which it is 180 degrees from the sun, and its disk appears fully illuminated when viewed from earth.

Fusion (Thermonuclear). The joining together of lighter atomic nuclei to form heavier ones, accompanied by a transformation of matter into energy. This is the basic means of energy production in stars.

Galactic equator. A great circle on the celestial sphere representing the mid-line of the Milky Way.

Galactic poles. The two points on the celestial sphere that are located 90 degrees from the galactic equator.

Galactic star cluster. Another name for open star clusters, which may contain up to several hundred stars in irregularly shaped groups located in the disk of the Milky Way.

galaxy. A gravitationally bound assembly of between several million to hundreds of billions of stars.

Galaxy. The name of the galaxy that contains the solar system; also, Milky Way Galaxy.

Giant star. A star in late stages of its evolution, having an inflated diameter and a high luminosity.

Globular cluster. A gravitationally bound, spherical group of hundreds of thousands of stars, which orbits the center of the galaxy in a spherical volume of space known as the galactic halo. Stars in these clusters are believed to have been formed in the earliest days of the galaxy's history.

Globule. A small, dense concentration of dust and gas, possibly a site where new stars may form.

Gravitation. A property by which matter is attracted to itself.

Gravitational collapse. An unchecked condensation of matter produced by the influence of gravity.

Gravitational energy. Energy released when matter in a system becomes more concentrated.

Gravitational lens. A phenomenon in which the gravitational field of a celestial object such as a galaxy acts as a lens that focuses the image of a more distant object viewed along nearly the same line of sight.

Great circle. A circle on a spherical surface in which the plane of the circle passes through the center of the sphere.

H I region. An interstellar cloud of cool hydrogen gas at a temperature of less than 100°K.

H II region. An expanding region of interstellar hydrogen surrounding hot, young stars heated to temperatures of about 10,000°K by ultraviolet radiation from them. H II regions are also known as Strömgren spheres.

Halo (of a galaxy). A spherical region surrounding the disk of a spiral galaxy. The halo is thinly populated by stars that orbit the galactic center either singly or as members of globular clusters.

Hayashi lines. Theoretical paths of protostars on the H-R diagram as they approach the main sequence.

Heavy elements. In general astronomical usage, all chemical elements other than hydrogen and helium.

Herbig-Haro objects. Patches of nebulosity believed by some astronomers to be formed as jets of material ejected by protostars collide with interstellar material near the surface of a molecular cloud.

Hertzsprung-Russell (H-R) diagram. A chart used to compare the absolute magnitudes and spectral classes of stars.

High-velocity star. A star whose velocity differs substantially from that of the sun. These stars are generally temporary visitors to the sun's vicinity in the galactic disk and are normally located in the halo region of the galaxy.

Horizon. The astronomical horizon is defined as a great circle on the celestial sphere, located 90 degrees from the zenith.

Hour circle. One of the great circles on the celestial sphere that pass through the celestial poles and that are used to mark hours of right ascension in the equatorial coordinate system.

Hubble constant. A number used to correlate the redshift velocity of celestial objects with their distances from earth. Most estimates of the Hubble constant lie between 50 and 100 kilometers per second per million parsecs.

Hubble law. A statement of the relationship between redshifts and distances from earth for remote objects.

Hubble time. A period of 10 to 20 billion years, which represents the time required for the universe to have expanded from the primeval fireball to reach its present estimated size.

Infrared radiation. A form of electromagnetic radiation with wavelengths in the range between those of red visible light and microwaves. Infrared radiation is also known as radiant heat.

Interferometer. A device designed to blend two or more beams of electromagnetic radiation from a single source in order to provide increased resolution and more information about the object's angular size.

Interstellar dust. Grains of matter of about 0.0002 millimeter in diameter (about the size of particles of very fine clay) that obscure the light from background stars.

Interstellar gas. Gas, mostly consisting of hydrogen, found in regions between stars.

Interstellar medium. Gas, dust, magnetic fields, and cosmic rays found between the stars.

Irregular variable. A variable star with unpredictable periods of light variations.

Kelvin temperature scale. A temperature scale whose units are of the same size as Celsius degrees, and whose zero point is equal to −273.16 degrees Celsius or −459.69° Fahrenheit.

Kepler's laws. Three laws devised by Johannes Kepler in the early seventeenth century to describe the orbital motions of planets.

Kinetic energy. The energy of motion.

Light. The wavelengths of electromagnetic radiation that are visible to the human eye.

Light curve. Characteristic changes in the brightness of a variable star over a period of time.

Light-year. The distance that light travels during one year through a vacuum. A light-year is equal to about 9.5 trillion kilometers, or 6 trillion miles.

Limiting magnitude. The magnitude of the faintest object visible through a certain telescope or under specific seeing conditions.

Local Group of Galaxies. A group of about twenty gravitationally bound galaxies, including the Milky Way, the Great Galaxy in Andromeda, and M33 in the constellation Triangulum.

Local Supercluster. A vast collection of many clusters of galaxies, including the Local Group and the Virgo Cluster.

Luminosity. The intrinsic brightness of a star.

Luminosity class. A category used to describe a star's luminosity and position on the H-R diagram.

Main sequence. The region on the H-R diagram that represents the luminosity and spectral class of about 90 percent of the stars in the neighborhood of the sun. During its main-sequence stage, a star derives its energy from the thermonuclear fusion of hydrogen to helium in its core.

Mass. A measure of the amount of material contained in an object.

Mass ejection. A process by which matter is ejected into space from the surface of a star.

Mass-luminosity relation. A graph that relates the mass and luminosity of main-sequence stars.

Meridian (celestial). A great circle on the celestial sphere that passes through an observer's zenith and north and south horizon points. The time of meridian transit represents a celestial object's highest elevation in the sky.

Messier's catalogue. A listing of star clusters and nebulae prepared by Charles Messier during the late eighteenth century. These objects are often known by their "M" numbers, from this catalogue.

Meteor. Often called a "shooting star," this phenomenon is an incandescent trail of gas emitted when a piece of rock or iron is heated by friction as it plunges through the earth's atmosphere.

Meteor shower. An encounter between the earth and a swarm of meteoroids that results in many meteors appearing to radiate from near one point in the sky.

Meteorite. A piece of rock or stone that survives its plunge through the atmosphere as a meteor and reaches the earth's surface.

Meteroid. A small piece of rock or stone in orbit around the sun.

Microwave. A form of very short wave radio energy, located on the electromagnetic spectrum just beyond the infrared.

Milky Way. The faintly glowing band of light that circles the sky and represents our view of stars and nebulae in the disk of our galaxy. "Milky Way" is also used as the proper name of this galaxy.

Minute of arc. A minute of arc is equal to 1/60 of an angular degree or to 60 arcseconds.

Mira-type variable. A type of long-period, irregular variable stars, named after their prototype, Mira, also known as Omicron Ceti.

Molecular cloud. A cloud in one of the Milky Way's spiral arms that consists mainly of molecular hydrogen at temperatures of about 10 to 50 degrees Kelvin. The molecular clouds also contain traces of other gases, such as carbon monoxide, and quantities of interstellar dust. It is believed that these clouds are sites for the formation of new stars.

Molecule. The smallest particle of a chemical compound that retains the properties of the compound.

Nebula. A cloud of gas or dust in interstellar space.

Neutron star. A supernova remnant of extremely high density comprised almost entirely of neutrons. Some neutron stars manifest themselves as pulsars.

New General Catalogue (NGC). A catalogue of star clusters, nebulae, and galaxies.

Newton's laws. The laws of gravitation and mechanics devised by Isaac Newton in the late seventeenth century.

Nova. A star that suddenly increases up to several hundreds of thousands of times in brightness. Novae result when hydrogen streaming from the giant component of a close binary system accumulates on its white dwarf companion in sufficient amounts to trigger a thermonuclear explosion. A great burst of heat and light energy is unleashed, and the acquired mass is blown away from the surface of the white dwarf.

Nucleosynthesis. The processes within stars whereby atomic nuclei are assembled from lighter nuclear components as a result of thermonuclear fusion.

OB association. A group of up to about one hundred brilliant, young stars formed near each other in spiral arms of the Milky Way and other galaxies. An association contains too few stars to be gravitationally stable, and the group eventually disperses after a few million years.

Objective lenses. The lenses of refracting telescopes that collect light and form images of objects that may be magnified by eyepiece lenses.

Obscuration (interstellar). The dimming of light from celestial objects by the intervention of interstellar dust.

Occultation. An eclipse of one celestial object by another.

Open cluster (galactic cluster). A group of up to several hundred stars found in the disk and particularly in the spiral arms of the galaxy. Open star clusters are gravitationally bound and their member stars remain together in irregularly shaped groups.

Optical double. These are double stars that only appear to be close to each other but are not parts of the same, gravitationally bound system. Optical doubles are seen when stars that are at different distances from earth are viewed along nearly the same line of sight.

Parallax (stellar). A slight variation in the apparent position of a star on the celestial sphere caused by the change in an observer's vantage point as earth moves to a new location along its orbit. Measurements of stellar parallax enable astronomers to determine the distances of stars that lie less than about 120 light-years from earth. The trigonometric parallax of a star is equal to the angular distance between the earth and the sun, as observed from the star's distance. All stars beyond the sun have trigonometric parallaxes of less than one arcsecond. It is the basic method for determining distances of stars.

Parsec. The unit of space distances preferred by astronomers. A parsec is defined as the distance from earth at which a star would have a parallax of one arcsecond. This is equal to about 3.26 light-years.

Period. The interval of time between successive occurrences of a particular event.

Period-luminosity relation. The relationship between the luminosities (intrinsic brightnesses, absolute magnitudes) and periods of light variation in Cepheid-variable stars. Cepheids with longer periods have higher luminosities than Cepheids with shorter periods. The period-luminosity relation is used to help estimate distances in space.

Photometry. An astronomical method for measuring the intensities of various wavelengths of the electromagnetic spectrum of celestial objects.

Planet. One of nine celestial objects that orbit the sun and shine by reflected sunlight. Other stars may also have planetary systems.

Planetary nebula. A ring of luminous gas surrounding certain stars that are near the end of their evolution. These nebulae superficially resemble planets but are actually our cross-sectional view of an expanding shell of gas that has been ejected from the old stars prior to their transformation into white dwarfs.

Polar axis. Earth's axis of rotation, or the axis of a telescope that is aligned with that of the earth.

Population I stars. Relatively young stars located in a galaxy's disk and especially in its spiral arms. The sun, with an age of about 4.6 billion years, is one of the older population I stars. These stars have been enriched by quantities of metals and other heavy elements formed during episodes of nucleosynthesis that occurred during the evolution of previous generations of stars in the galaxy's disk. The substance and activities of life on the earth are made possible due to the presence of heavy elements such as carbon, oxygen, and iron.

Population II stars. These stars are usually found in a galaxy's halo or in its central bulge. Occasionally the orbit of a halo star carries it through the galaxy's disk, as is the case with the star Arcturus. Population II stars are survivors from the early stages in the formation of galaxies, and as a result, they never had an opportunity to be enriched by significant quantities of heavy elements spewed into space by aging stars and incorporated into new stellar generations within a galaxy's disk.

Position angle. The angular distance measured from north through east on the celestial sphere that defines the position of a secondary star with respect to the primary in a double star system.

Precession. The conical wobble of earth's axis of rotation over a period of about 26,000 years, caused by gravitational influences on earth from the sun, moon, and other planets of the solar system. One effect of precession is a slow change in the positions of stars relative to earth-oriented coordinate systems.

Primeval fireball. The explosion that, according to the Big Bang theory, formed our expanding universe.

Proper motion. The angular velocity of a star's actual motion across our line of sight, on the celestial sphere.

Protostar. A contracting region of gas and dust that may lead to the formation of a new star.

Pulsar. A supernova remnant that is a source of rapidly pulsating radio emissions with periods of a few seconds and less. These objects are manifestations of neutron stars that have diameters of about the same size as that of the earth.

Pulsating variable stars. Stars that vary in both luminosity and diameter over a period of time.

Quasar. These objects appear to have starlike centers and are slightly variable in brightness, with periods of a few years or less. They are also characterized by large redshifts, which have been interpreted by many astronomers to indicate distances from earth in the order of billions of light-years. The estimated distances, sizes, and energy production of quasars have not yet been fully explained.

Radial velocity. The component of a star's space motion directly towards or away from the earth. A positive (+) radial velocity means the object is moving away from earth and negative (−) implies motion towards the earth. Radial velocities are determined by measurements of Doppler shifts in the spectra of celestial objects.

Radiation. A way in which energy is transmitted across space.

Radio astronomy. The branch of astronomy that deals most directly with observations of electromagnetic energy having wavelengths of one millimeter or longer emitted by celestial objects.

Radio telescope. A device used to collect and detect radio energy emitted by celestial objects.

Recurrent nova. A nova whose eruption has been observed on more than one occasion.

Red giant. A star in the late stages of its evolution, whose luminosity and color place it on the upper-right-hand portion of the H-R diagram, above the main sequence. Although red is the most intense color produced by these stars, the blend of all colors radiated by red giants causes these stars to appear yellow or yellow-orange to our eyes.

Redshift. A shift of absorption or emission lines towards the red end of an object's spectrum. Redshifts may be caused by either an object's motion away from the earth or by the effects of an intense gravitational field. Redshifts provide a means for estimating the distances of galaxies more than a few hundred million light-years from earth.

Reflecting telescope. A telescope that uses a concave mirror to collect light and to form images.

Reflection nebula. A cloud of interstellar dust visible because of its reflection of light from neighboring stars.

Refracting telescope. A type of telescope that uses a lens or set of lenses to collect light and form an image.

Resolution. The ability of a telescope or other instrument to detect the separations of double stars or to show fine details of an object.

Revolution. The motion of one object around another.

Right ascension. A numbering system for equatorial coordinates used to indicate the east-west position of an object on the celestial sphere. The coordinates of right ascension range from zero through 23 hours and back to zero, a span of 24 hours. Each hour of right ascension is divided into minutes and seconds. The minutes (m) and seconds (s) of right ascension are related to times and are not equal to the minutes (′) and seconds (″) of angular measure, which are used to subdivide degrees of declination. The great circle that indicates zero hours of right ascension passes through the celestial poles and the point of the vernal equinox on the celestial equator.

Rotation. The motion in which an object spins around an axis that passes through itself.

RR Lyrae stars. A type of pulsating variable star that has a period of less than one day and is usually found in the galactic halo or bulge. As is the case with Cepheid variables, the period-luminosity relation for RR Lyrae stars serves as an important tool for estimating distances in space. RR Lyrae stars are also known as cluster-type variables, since they are found in most globular clusters.

Schwarzschild radius. The radius at which a massive contracting object becomes a black hole. When this size is reached, light from the object is no longer able to escape the object's highly concentrated gravitational field.

Seeing conditions. The conditions in the atmosphere that affect our ability to see well-defined images through a telescope.

Separation (of double stars). The angular distance between double stars.

Shock wave. A shell of compressed gas that expands through a medium as a result of the passage of material through the medium at a speed greater than that of sound in that medium.

Sidereal day. The time interval between two successive transits of the celestial meridian by the vernal equinox.

Sidereal time. The number of hours, minutes, and seconds since the vernal equinox last crossed the celestial meridian. This is equivalent to the right ascension of an object currently on the observer's meridian.

Singularity. A point at the center of a black hole where the known laws of physics apparently no longer hold true.

Sky Screen. A region of sky, extending from north through east to south, approximately 30° above the horizon, where stars featured in *The Star Guide* are presented on specific dates throughout the year at about 9:00 P.M., local time.

Small circle. A circle on a spherical surface, defined by the intersection of the sphere with a plane that does not pass through its center.

Solar antapex. The apparent point on the celestial sphere that the solar system is moving away from, relative to neighboring stars.

Solar apex. The point on the celestial sphere towards which the sun is apparently traveling, relative to other stars in its neighborhood.

Solar system. The gravitationally bound system that includes the sun, planets, satellites, asteroids, comets, and meteoroids.

Space motion. The total velocity of a star as it moves through space relative to the sun.

Space telescope (Hubble). A reflecting telescope with a 2.4-meter objective mirror, which is expected to revolutionize the quality of many astronomical observations as it orbits the earth, at an altitude of about 500 kilometers.

Spectral class. A classification of stars made on the basis of comparisons of various spectral lines. The sequence of spectral classes is designated O, B, A, F, G, K, M, an order that represents decreasing surface temperatures.

Spectroscope. A device used to analyze light.

Spectroscopic analysis. The study of a spectrum in order to learn about its source of light.

Spectroscopic parallax. An estimate of stellar distance based on a comparison of a star's apparent magnitude with the absolute magnitude characteristic of its spectral and luminosity classes.

Spectrum. An array of colors and/or other manifestations of electromagnetic radiation.

Spiral arms. Regions of gas, dust, and young stars that curve outwards from the centers of spiral galaxies.

Spiral galaxy. A type of galaxy characterized by a flattened, rotating disk of stars, which contains a system of spiral arms.

Star. A hot sphere of gas that shines with its own light.

Star cluster. A group of stars whose members are gravitationally bound to each other.

Stellar evolution. The changes which take place as stars form, mature, and expire.

Stellar wind. A flow of material from a star's surface out into space.

STR (Star Time Reference) numbers. A sequence of numbers, used in *The Star Guide,* to help determine the relative arrival times of stars at their Sky Screen positions.

Strömgren sphere. A region of heated and expanding hydrogen which surrounds an extremely hot and luminous young star or stars. Also called H II regions.

Subgiant. The first luminosity class attained by a star as it expands and evolves away from the main sequence to become a giant.

Summer solstice. The most northerly position of the sun on the ecliptic. The sun's arrival at this point, on or about June 21, marks the beginning of summer in the northern hemisphere.

Sun. The nearest star to earth.

Supercluster. A group of many individual clusters of galaxies extending over a region of space having a diameter on the order of 200 million light-years.

Supergiant. A star of exceptionally high luminosity and diameter.

Supernova. A cataclysmic explosion in which a star of high mass blows itself apart and in the process increases in luminosity up to a billion times. If the star's core survives, it contracts to become a neutron star or a black hole.

Supernova remnant. A neutron star, a black hole, or an expanding nebula, formed from the material ejected into space by a supernova.

T Tauri stars. A type of irregular variable star believed to be still in the process of pre–main-sequence gravitational contraction.

Telescope. An instrument used to collect and magnify electromagnetic radiation from distant objects.

Transit. The passage of a celestial object across the celestial meridian.

Ultraviolet radiation. A form of electromagnetic radiation having wavelengths just shorter than violet.

Universe. The entirety of all known matter and energy.

Variable star. A star that varies in brightness.

Vernal equinox. The point at which the ecliptic intersects the celestial equator and which is the location of the sun on or about March 21. The sun's arrival at the vernal equinox marks the beginning of spring in the northern hemisphere.

Very Large Array (VLA). A group of twenty-seven radio telescopes, each with a dish antenna having a diameter of 25 meters, located along three 21-kilometer-long tracks in New Mexico. When these telescopes are used in conjunction with each other, high-resolution radio images of celestial radio sources may be obtained.

Visual double star. A double star whose components have sufficient angular separation so that they may be seen as double when viewed through a telescope.

Wavelength. The distance between comparable points of an electromagnetic wave.

White dwarfs. The dense, collapsed remnants of stars near the end of their evolution.

Winter solstice. The most southerly point on the ecliptic, reached by the sun on or about December 21. This event marks the start of winter in the northern hemisphere.

Zenith. The point directly over an observer, which is located 90 degrees from the horizon.

Zero-age main sequence. The position on an H-R diagram occupied by stars that have just arrived on the main sequence, after having ceased their gravitational contractions.

Zero hour of right ascension. Defined by the semicircle on the celestial sphere passing through the vernal equinox as well as the celestial poles, this coordinate represents the starting line for measuring right ascension.

Zodiac. A region on the celestial sphere that extends eight degrees on both sides of the ecliptic for a total width of 16 degrees. The zodiac is where the moon and planets may be seen.

Bibliography

Sky Observations

Allen, Richard Hinckley, *Star Names: Their Lore and Meaning*. New York: Dover Publications, 1963.

Burnham, Robert, Jr., *Burnham's Celestial Handbook*. 3 vols. New York: Dover Publications, 1978.

Chartrand, Mark R., III, *Skyguide*. Racine, WI: Golden Press, 1983.

Gallant, Roy A., *The Constellations*. New York: Four Winds Press, 1979.

Kunitzsch, Paul, *Arabische Sternnamen in Europe*. Wiesbaden, W. Germany: O. Harrossowitz, 1959.

Mallas, John H., and Kreimer, Evered, *The Messier Album*. Cambridge, MA: Sky Publishing, 1978.

Martin, R. Newton, and Menzel, Donald Howard, *The Friendly Stars*. New York: Dover Publications, 1964.

Mayall, R. Newton, and Mayall, Margaret W., *Olcott's Field Book of the Skies*. 4th ed. New York: G. P. Putnam's Sons, 1954.

Mayall, R. Newton; Mayall, Margaret W.; and Wyckoff, J., *Sky Observer's Guide*. New York: Golden Press, 1977.

Muirden, James, *The Amateur Astronomer's Handbook*. 3d ed. New York: Harper & Row, 1982.

Mullaney, James, and McCall, Wallace, *The Finest Deep-Sky Objects*. Cambridge, MA: Sky Publishing, 1978.

Ottewell, Guy, *Astronomical Calendar*. Annual. Greenville, SC: Guy Ottewell, Furman University.

———, *Astronomical Companion*. Greenville, SC: Guy Ottewell, Furman University.

Ray, H., *The Stars: A New Way to See Them*. 3d ed. Boston: Houghton Mifflin, 1967.

Robinson, Leif J., ed., *The "Sky and Telescope" Guide to the Heavens*. Cambridge, MA: Sky Publishing, 1980.

Royal Astronomical Society of Canada, *Observer's Handbook*. Annual. Toronto: University of Toronto Press.

Sherrod, P. Clay, *A Complete Manual of Amateur Astronomy*. Englewood Cliffs, NJ: Prentice-Hall, 1981.

U.S. Naval Observatory, *The Astronomical Almanac*. Annual. Washington, D C: U.S. Government Printing Office.

Star Atlases

Edmund MAG 5 Star Atlas. Barrington, NJ: Edmund Scientific, 1974.

Norton, W., *Sky Atlas*. Cambridge, MA: Sky Publishing, 1971.

Tirion, Wil, *Sky Atlas 2000.0*. Cambridge, MA: Sky Publishing, 1981.

Textbooks

Abell, George O., *Exploration of the Universe.* 4th ed. Philadelphia: Saunders College Publishing, 1982.

——, *Realm of the Universe.* 2d ed. Philadelphia: Saunders College Publishing, 1980.

Berman, Louis, and Evans, John C., *Exploring the Cosmos.* 4th ed. Boston: Little, Brown, 1983.

Brandt, John C., and Maran, Stephen P., eds., *New Horizons in Astronomy.* San Francisco: W. H. Freeman, 1979.

Hoyle, Fred, *Astronomy and Cosmology: A Modern Course.* San Francisco: W. H. Freeman, 1975.

Pasachoff, Jay M., *Contemporary Astronomy.* 2d ed. Philadelphia: Saunders College Publishing, 1981.

Motz, Lloyd, and Duveen, Anneta, *Essentials of Astronomy.* 2d ed. New York: Columbia University Press, 1977.

Zeilik, Michael, *Astronomy: The Evolving Universe.* 3d ed. New York: Harper & Row, 1982.

Various Aspects of Astronomy

Ashbrook, Joseph, *The Astronomical Scrapbook: Skywatchers, Pioneers, and Seekers in Astronomy;* edited by Leif J. Robinson. Cambridge, MA: Sky Publishing Corp, 1985.

Asimov, Isaac. *The Universe: From Flat Earth to Quasar.* New York: Discus, Avon, 1976.

Bok, Bart J., and Bok, Priscilla F., *The Milky Way.* 5th ed. Cambridge, MA: Harvard University Press, 1981.

Brandt, John C., and Maran, Stephen P., eds., *The New Astronomy and Space Reader.* San Francisco: W. H. Freeman, 1977.

Einstein, Albert, *Relativity: The Special and General Theory.* New York: Crown, 1961.

Ferris, Timothy, *Galaxies.* San Francisco: Sierra Club Books, 1980.

——, *The Red Limit.* New York: William Morrow, 1977.

Gallant, Roy A., *National Geographic Picture Atlas of Our Universe.* Washington, DC: National Geographic Society, 1980.

Gingerich, Owen, ed., *New Frontiers in Astronomy.* San Francisco: W. H. Freeman and Co., 1975.

Gingerich, Owen, ed., *Cosmology + 1: Readings from Scientific American.* San Francisco: W. H. Freeman and Co., 1977

Hopkins, Jeanne, *Glossary of Astronomy and Astrophysics.* 2d ed. Chicago: University of Chicago Press, 1980.

Jastrow, Robert, *Red Giants and White Dwarfs.* New York: Harper & Row, 1967.

Kaufmann, William J., III, *The Cosmic Frontiers of General Relativity.* Boston: Little, Brown, 1977.

King, Henry C., *The History of the Telescope.* New York: Dover Publications, 1979.

Kippenhahn, Rudolf, *100 Billion Suns: The Birth, Life, and Death of the Stars.* New York: Basic Books, 1983.

Kirby-Smith, Henry Tompkins, *U.S. Observatories: A Directory and Travel Guide.* New York: Van Nostrand Reinhold, 1976.

Malin, David, and Murdin, Paul. *Colours of the Stars.* Cambridge: Cambridge University Press, 1984.

McGraw-Hill Encyclopedia of Astronomy. New York: McGraw-Hill, 1983.

Peltier, Leslie C., *Starlight Nights: The Adventures of a Star-Gazer.* Cambridge, MA: Sky Publishing, 1980.

Proctor, Percy M., *Star Myths and Stories.* New York: Exposition, 1972.

Sagan, Carl, *The Cosmic Connection: An Extraterrestrial Perspective.* New York: Anchor Press–Doubleday, 1980.

———, *Cosmos.* New York: Random House, 1980.

Shapley, Harlow, *Galaxies.* Cambridge, MA: Harvard University Press, 1972.

Shipman, Harry L., *Black Holes, Quasars, and the Universe.* 2d. ed. Boston: Houghton Mifflin, 1980.

Sullivan, Walter, *Black Holes.* New York: Anchor Press, 1979.

Warner, Deborah Jean, *Alvan Clark and Sons: Artists in Optics.* Washington, DC: Smithsonian Institution Press, 1969.

———, *The Sky Explored: Celestial Cartography 1500–1800.* New York: A. R. Liss, 1979.

Weinberg, S., *The First Three Minutes.* New York: Basic Books, 1977.

Willard, Berton C., *Russell W. Porter: Arctic Explorer, Artist, Telescope Maker.* Freeport, ME: Bond Wheelwright, 1976.

PERIODICALS

Astronomy

Astronomy, AstroMedia Corp., P.O. Box 92788, Milwaukee, WI 53202

Mercury, Journal of the Astronomical Society of the Pacific, 390 Ashton Avenue, San Francisco, CA 94112

Sky Calendar, Abrams Planetarium, Michigan State University, East Lansing, MI 48824 (newsletter)

Sky and Telescope, 49 Bay State Rd., Cambridge, MA 02238-1290

Science in General, with Some Articles about Astronomy

Discover, Time, Inc., 3435 Wilshire Blvd., Los Angeles, CA 90010

Natural History, American Museum of Natural History, Central Park West at 79th St., New York, NY 10024

Omni, 909 Third Avenue, New York, NY 10017

Science, American Association for the Advancement of Science, 1515 Massachusetts Ave. N.W., Washington, DC 20005

Science News, 231 West Center St., Marion, OH 43302

Scientific American, 415 Madison Avenue, New York, NY 10017

Smithsonian Magazine, Smithsonian Institution, Washington, DC 20560

Index